T0136778

SOLAR RADIATION AND CLOUDS

While we have written many words and calculated many numbers to describe the interplay of light and clouds, perhaps our entire effort is not as significant as the words of John Muir

> "Light. I know not a single word fine enough
> for light . . . holy, beamless, bodiless,
> inaudible floods of light".[1]

---

[1]From *John Muir's Wild America*, by Tom Melham, published by the National Geographic Society, 1976.

METEOROLOGICAL MONOGRAPHS

VOLUME 17 | MAY 1980 | NUMBER 39

SOLAR RADIATION AND CLOUDS

by
Ronald M. Welch, Stephen K. Cox and John M. Davis

Published by the American Meteorological Society
45 Beacon St., Boston, MA 02108

ISBN 0-933876-49-1
ISSN 0065-9401

American Meteorological Society
45 Beacon Street, Boston, MA 02108

Printed in the United States of America
by Lancaster Press, Lancaster, PA

Acknowledgments

We gratefully acknowledge the contributions of Ms. Pauline Martin and Sandy Wunch in the preparation phases of this manuscript, for without their dedication and untiring efforts this monograph would not have been possible. We also wish to thank the scientific reviewers of this monograph for a thorough and constructive reading of the manuscript.

Many of the computations were made utilizing the facilities of the National Center for Atmospheric Research Computing Center which is supported by the National Science Foundation. This research has been conducted under the auspices of a sequence of grants from the National Science Foundation and the NOAA GATE Project Office.

TABLE OF CONTENTS

LIST OF FIGURES

LIST OF TABLES

Preface

The research reported in this monograph has evolved in a very pragmatic fashion. It represents an attempt to explore the extreme values as well as the more probable values of the radiative characteristics of water and ice clouds in the solar wavelengths. Since the mid 1960's a number of observations of shortwave absorption in clouds have been reported which appear to be larger than classical theory can explain. One of the authors of this monograph (SKC) has been closely associated with the collection of a number of these observations of solar fractional absorption. While we freely admit that the observations do contain some error on occasion, the differences between the observations and theoretical estimates are so large that they may not be dismissed as observational error alone; therefore we have explored several plausible explanations for what have heretofore been labeled excessive values of solar absorption.

We begin with a review of results based on commonly used classical theory in order to define the present state of understanding of the shortwave radiative characteristics of clouds. We then proceed to investigate the effects of bimodal droplet size distributions and, in the case of ice crystals, the implications of resonance suppression theory. Next, the impacts of non-homogeneous microphysical characteristics are explored; both radiative characteristics and observational sampling problems caused by the nonuniform clouds are examined. We have not successfully isolated the explanation for the large values of absorption deduced from observations, but we do present results of detailed calculations which for the first time show that these large values are theoretically possible.

The above investigations generated a large data base of cloud radiative characteristics which we used to seek a plausible answer for the disagreement between observations and theory. Since the effort to produce these data was considerable, we have elected to tabulate many of the results and include them in the text for other investigators to refer to and use. While the inclusion of so many tables in the text makes the monograph less readable, it is our opinion that these somewhat cumbersome tabulations make the monograph potentially much more useful.

S. K. Cox
J. M. Davis
R. M. Welch
Colorado State University

CHAPTER 1

Introduction

It has long been recognized that clouds play a crucial role in the determination of the earth's energy budget. The spatial distribution of radiative heating is the fundamental energy source responsible for driving the dynamics of both atmosphere and oceans. Energy budget studies show that both the solar and infrared radiation fields are significantly altered by cloud cover, thereby affecting atmospheric heating/cooling profiles and surface temperature (Paltridge, 1974b). Furthermore, recent measurements by Cox *et al.* (1973) and Reynolds *et al.* (1975) show that direct absorption of solar radiation in clouds is an important tropospheric heat source.

While most dynamical models have either ignored radiation forcing terms altogether or parameterized the radiational heating/cooling rates in a very approximate fashion, recent measurements and models have demonstrated that radiation-dynamical interactions play a direct role concerning cloud and precipitation development, cyclogenesis and atmospheric stability. A number of these studies are summarized below.

Paltridge (1974a) reported that radiational heating can be a dominant factor in determining thickness, water content and the general character of stratocumulus clouds. These observations also showed that reduction of shortwave absorption, treated as increased infrared cooling, tended to dissipate such clouds.

Lilly (1968) developed a model of a cloud-topped mixed layer corresponding to stratocumulus clouds. Through the use of this model he exhibited the importance of the radiative budget at the top of the cloud in maintaining the cloud-top mixed layer system. Schubert (1976) and Schubert *et al.* (1977) have expanded Lilly's model and have shown that for the case of the stratocumulus cloud deck topped by a strong inversion, radiative divergence in the vicinity of the cloud top modulates the strength of mixing in the subcloud layer as well as the cloud thickness and cloud top height. Schubert also showed significant diurnal variability related to in-cloud solar heating, compensating for the longwave cooling.

Albrecht (1979) has shown the interaction between the time-dependent temperature and moisture profiles and radiative divergence for the undisturbed trade wind layer in the tropics. Albrecht was quite successful in modeling the evolution of the boundary layer temperature and moisture structure and its diurnal variability.

Cox (1969) found that radiative components are of the same order of magnitude as latent and sensible heat terms and may significantly influence synoptic-scale development. Radiative processes were shown to influence vorticity changes for the nascent cyclone and to contribute 10–33% of the total thickness change.

Yanai *et al.* (1972) attempted to determine the interaction between large-scale sensible, latent and radiational heating and cooling rates and tropical waves or cloud clusters in order to obtain a measure of the activity of cumulus convection. This study showed a close relationship between diurnal radiation balance variations and tropical precipitation patterns.

An infrared study by Albrecht and Cox (1975) applied realistic radiative forcing to a tropical synoptic-scale numerical model first described by Holton (1971). In this paper Albrecht and Cox showed very significant forcing from the radiative divergence patterns alone, although other energy budget components were included in the study. They further reported a strong dependence of the diagnosed motion field upon the position of the cloud-induced radiative divergence to the convective heating in the wave disturbance.

Reynolds *et al.* (1975) showed that differential vertical heating by shortwave radiation in low clouds acts to destabilize the entire layer, thus potentially enhancing cloud growth. In contrast, very strong heating rates observed in high cirrus clouds not only warm the upper troposphere and inhibit warming at lower levels, but also tend to stabilize the entire atmospheric air column (Griffith *et al.*, 1980).

Large differential heating between cloud-covered and clear air regions may lead to enhanced horizontal temperature and moisture gradients, thereby generating a pressure gradient to drive local atmospheric motions. Gray and Jacobson (1977) conclude that radiational heating differences between cloud and cloud-free areas, and not condensation forcing, drive deep layer convergence and are the main mechanism for cloud system maintenance and early intensification. Observational evidence by Gray and Jacobson (1977) and Foltz and Gray (1979) tend to confirm the hypothesis of Yanai *et al.* (1973) that radiational differences between cloud regions and surrounding cloud-free regions of tropical systems are the fundamental driving mechanism for large observed morning deep cumulus convection maxima and associated precipitation patterns.

An analysis of the GARP Atlantic Tropical Experiment (GATE) data by Grube (1977) shows that upper level warmings extending through a layer 200–300 mb thick occur in clear air adjacent to regions of deep convection and coincide with the time of daily solar heating maxima. Warming of the upper troposphere is a requirement for the development of tropical cyclones. It is interesting that such warmings were not observed on clear days or days characterized by suppressed convective activity. A theoretical model by Fingerhut (1977) shows that radiation is fundamental to the maintenance of cloud cluster disturbances. Frank (1977) concludes that for regions of deep convective activity within the GATE A/B and B scale arrays, passage of easterly wave troughs may act as a trigger mechanism for convection. These investigations have concluded that upper level warmings are too large to be due solely to the effects of daily solar or infrared heating in clouds. The present paper concentrates upon solar radiation and will demonstrate that solar heating rates can be very large and strongly height dependent.

Absorption in clouds and, therefore, cloud heating rates are dependent upon many variables: cloud thickness, cloud top height, liquid water content, drop size distribution, atmospheric water vapor profile, cloud dimensions and solar zenith angle. Radiation measurements by Yamamoto (1962), Roach (1961), Cox *et al.* (1973) and Reynolds *et al.* (1975) have improved our understanding of radiation parameters used in theoretical radiation models dealing with solar energy absorption and atmospheric heating rates. Previous theoretical calculations by Zdunkowski and Korb (1974), Twomey (1976), Welch *et al.* (1976) and Feigelson and Krasnokutskaya (1978) have shown that radiative heating/cooling rates are highly variable both within individual cloud boundaries and from one cloud type to another.

However, previous theoretical calculations of solar radiation absorption provide maximum values of 15–20% while absorptions as large as 30–52% are observed (Reynolds *et al.* 1975). As pointed out by Twomey (1976), such large absorptions require either total absorption for wavelengths greater than 0.7 μm or a sizeable degree of absorption at shorter wavelengths. Twomey concludes that absorption by pure water alone cannot be responsible for such large absorption values, while Twomey (1972) showed that enhancement of particulate absorption as a result of scattering in clouds cannot provide such large values. The present set of investigations attempts a thorough study of radiation absorption in clouds in order to provide a solution to this perplexing question.

Each chapter of this monograph deals with a particular aspect of absorption of solar radiation in clouds. Chapter 2 considers cloud heating rates as a function of cloud thickness, cloud top height, liquid water content, solar zenith angle and drop size distributions. In addition, it considers average daily cloud heating rates and the effect of water vapor absorption above the cloud top. Chapter 3 considers the anomalously large absorption rates observed by Reynolds *et al.* (1975) and shows that such very large absorptions may be the result of a bimodal drop size distribution. Comparisons with monomodal small drop size distributions and monomodal large drop size distributions are presented. Chapter 4 considers the effect of ice clouds for both monomodal and bimodal size distribution functions. Nonspherical corrections are applied to classical Mie theory and are shown to be very significant for small ice crystals and negligible for large ice crystals. Furthermore, it is shown that the size distribution of small ice crystals is a more important determinant of the solar radiation field in ice clouds than the large ice crystals. Calculations for both monomodal and bimodal ice crystal size distributions are given and compared with the observations reported by Griffith *et al.* (1980). Chapter 5 extends the results of the previous three parts to finite clouds, adding to the work of Davis *et al.* (1979a,b). Cloud radiative characteristics are examined for both monomodal and bimodal droplet size distributions in both water and ice clouds as a function of cloud thickness, cloud width, cloud top height and solar zenith angle. In addition, height variations of liquid water content and drop size distributions within clouds are examined. Finally, Chapter 6 considers the effect of horizontally inhomogeneous regions within finite clouds. The goal of this section is to determine how such horizontal inhomogeneities affect the cloud bulk radiative properties. Furthermore, selected "light channels" are modeled within the cloud. This is an important consideration since Platt (1976) has reported that clouds have horizontal variations in liquid water content, even though at a given height the droplet size distribution remained unchanged. In this section random distributions of various liquid water contents are distributed throughout the cloud.

CHAPTER 2

The Effect of Monomodal Drop Size Distributions, Cloud Top Heights, Cloud Thickness and Vertical Water Vapor Profiles upon Cloud Heating Rates and the Cloud Radiation Field

RONALD M. WELCH AND STEPHEN K. COX

2.1 *Attenuation data*

Attenuation of atmospheric radiation occurs through absorption and scattering by atmospheric gases, cloud droplets and particulate matter. In the present investigation the influence of dust on atmospheric and cloud radiation fields will be neglected.

2.1.1 GASEOUS ABSORPTION

We approximate water vapor transmittance through the use of a sum of exponentials

$$T = \sum_{i=1}^{N} b_i \exp(-K_i u_r), \qquad (2.1)$$

where u_r is the pressure-corrected water vapor path length, and b_i, K_i are empirical constants. The solar spectrum has been divided into eight regions. Seven of the regions comprise the water vapor bands described by Welch *et al.* (1976), while the eighth region is the sum of regions free from water vapor absorption. Rayleigh scattering has been neglected above the cloud in the present investigation and has negligible effect upon the parameters of interest within the cloud. Our primary emphasis is on the investigation of solar absorption in the wavelength region $\lambda > 0.70$ μm. Absorption by CO_2, ozone and trace gases has been neglected because it is small compared to water vapor and water droplet absorption in clouds. The water vapor absorption data are based upon the experimental studies of Howard *et al.* (1955). The empirical coefficients required in (2.1) are taken from Liou and Sasamori (1975).

2.1.2 DROPLET ATTENUATION PARAMETERS

Droplet absorption, scattering and extinction parameters, which are relatively slowly varying functions of wavelength, are computed from

$$\beta_\lambda = \int_0^\infty n(r)\, \pi\, r^2\, Q_\lambda(r)\, dr, \qquad (2.2)$$

where Q_λ is the wavelength-dependent Mie efficiency factor and $n(r)$ is the droplet size distribution function in terms of droplet radius r. Although Deirmendjian (1969, 1975) has given droplet attenuation parameters for several size distributions, the spectral regions chosen for the present investigation are incompatible with his selection of wavelengths. Therefore, the first section of the present study is based upon the attenuation parameters and phase function expansion coefficients given by Zdunkowski *et al.* (1967) which utilized the Best (1951) droplet distribution function.

Twomey (1976) and Wiscombe (1976) reported that the choice of the droplet distribution function is not critical as long as the distribution is relatively broad. The present investigation considers a variety of size distribution functions determined from observations in order to assess their influence on cloud absorption. Tampieri and Tomasi (1976) presented size spectra, based upon the measurements of Tverskoi (1965) which have been fit with the modified gamma distribution for clouds in various evolutionary stages. Size distributions of stratocumulus, stratus and nimbostratus clouds have been selected for the present study. Drop size distributions characteristic of cloud top and cloud base are also used in the present study. In addition, the C.5 and C.6 distribution functions given by Deirmendjian (1975) have been selected. These distributions are also based upon the modified gamma size distribution function

$$n(r) = a\, r^\alpha \exp\left[-\frac{\alpha}{\gamma}\left(\frac{r}{r_c}\right)^\gamma\right], \qquad (2.3)$$

expressed in cm^{-3} μm^{-1}; $n(r)$ is the number density of droplets with radius r and r_c is the modal radius of the distribution. The other parameters in (2.3) are empirically derived constants; typical values are given in Table 2.1, where

$$N = \int_0^\infty n(r)\, dr. \qquad (2.4)$$

Table 2.2 gives the extinction (β_e) and absorption (β_a) coefficients as a function of wavelength for the size distribution functions described above, along with the Best size distribution function at $w_L = 0.1$ and 0.2 g m^{-3}. The scattering parameter is the difference between the extinction and absorption parameters and is

TABLE 2.1. Cloud droplet size distribution parameters for stratocumulus, stratus, nimbostratus, C.5 and C.6 clouds based upon the modified gamma distribution function. (Notation, e.g., 2.8230-1 = 2.8230*10⁻¹).

Type	a	α	γ	r_c (μm)	w_L (g m^{-3})	N (cm^{-3})
Stratocumulus base	2.8230-1	5	1.19	5.33	0.141	10^2
Stratocumulus top	1.9779-1	2	2.46	10.19	0.796	10^2
Stratus base	9.7923-1	5	1.05	4.70	0.114	10^2
Stratus top	3.8180-1	3	1.30	6.75	0.379	10^2
Nimbostratus base	8.0606-2	5	1.24	6.41	0.235	10^2
Nimbostratus top	1.0969	1	2.41	9.67	1.034	10^2
C.5	0.5481	4	1.00	6.0	0.297	10^2
C.6	0.5000-4	2	1.00	20.0	0.025	10^{-1}

not shown. The complex index of refraction for water at these wavelengths is given by Zdunkowski *et al.* (1967) who weighted these values by the solar blackbody distribution function at 6000 K for each spectral interval. The computations were based upon the data compiled by Irvine and Pollack (1968) for wavelengths >0.7 μm. Values below 0.7 μm are taken from *Linkes Meteorologisches Taschenbuch* (1953). The present theoretical analysis using the delta-spherical harmonics technique is based upon the expansion of the phase function $P(\cos\theta)$ into a series of Legendre polynomials $P_\ell(\cos\theta)$ for each spectral interval,

$$P(\cos\theta) = \sum_{\ell=0}^{N} C_\ell P_\ell(\cos\theta). \tag{2.5}$$

Expansion coefficients C_1, C_2, C_3, C_4 are given for several droplet size distributions at selected wavelengths in Table 2.3. Coefficient C_0 is equal to unity in all cases.

The size distribution functions given in Tables 2.1–2.3 are quite diverse. Fig. 2.1 shows the drop size distribution functions for the cloud models reported by Tampieri and Tomasi (1976); in these data drop mode radius varies from approximately 5 μm to 10 μm. The

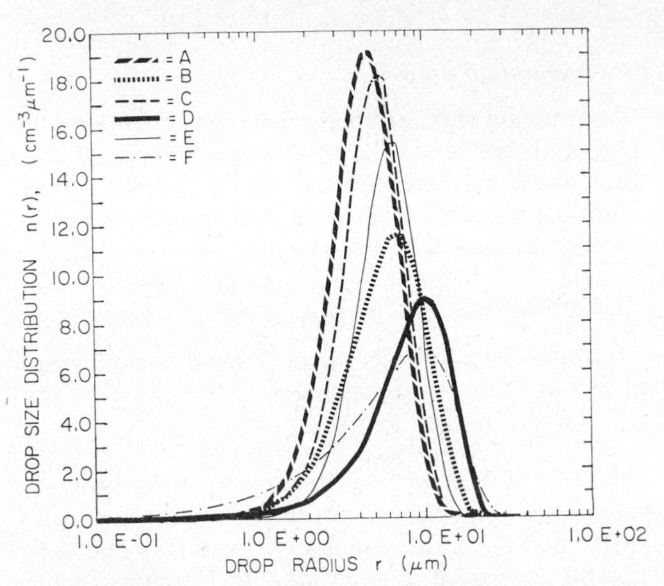

FIG. 2.1. Drop size distributions $n(r)$ for various cloud cases: A. stratus base, B. stratus top, C. stratocumulus base, D. stratocumulus top, E. nimbostratus base, F. nimbostratus top.

TABLE 2.2. Extinction (β_e) and absorption (β_a) coefficients (km^{-1}) for various droplet size distribution functions at selected wavelengths.

Type		Center of wavelength region (μm) 0.95	1.15	1.4	1.85	2.8	3.35	6.3
Stratocumulus base	β_e	29.02	29.44	29.80	30.20	31.34	31.69	34.92
	β_a	0.004	0.011	0.204	0.566	15.18	15.25	11.91
Stratocumulus top	β_e	90.13	91.13	91.59	92.90	94.58	95.83	105.1
	β_a	0.022	0.062	1.09	2.96	46.88	46.27	45.68
Nimbostratus base	β_e	40.82	41.22	41.68	42.30	43.41	44.05	50.38
	β_a	0.008	0.019	0.369	0.94	21.44	21.35	18.06
Nimbostratus top	β_e	101.8	102.8	103.2	104.6	106.4	107.7	115.9
	β_a	0.028	0.077	1.39	3.73	52.46	51.74	52.34
Stratus base	β_e	24.35	24.68	25.02	25.63	26.61	26.81	28.46
	β_a	0.0033	0.0095	0.162	0.45	12.66	12.80	9.64
Stratus top	β_e	53.51	54.06	54.49	55.23	56.54	57.32	64.39
	β_a	0.012	0.037	0.581	1.45	27.94	27.72	25.21
C.5	β_e	45.17	45.41	45.95	46.55	47.97	48.61	54.51
	β_a	0.008	0.028	0.406	1.12	23.61	23.49	20.69
C.6	β_e	0.774	0.775	0.779	0.783	0.789	0.794	0.815
	β_a	0.0006	0.0017	0.029	0.076	0.378	0.369	0.400
Best (w_L = 0.1)	β_e	35.03	35.63	36.39	38.01	38.45	39.68	37.73
	β_a	0.003	0.008	0.158	0.423	16.12	16.75	14.53
Best (w_L = 0.2)	β_e	46.25	46.85	47.60	49.11	50.01	51.16	53.89
	β_a	0.006	0.016	0.302	0.810	23.11	23.42	22.66

TABLE 2.3. Phase function expansion coefficients for several droplet size distribution functions at selected wavelengths.

Type		Center of wavelength region (μm)							
		0.765	0.95	1.15	1.40	1.85	2.8	3.35	6.3
Stratocumulus base	C_1	2.559	2.559	2.526	2.518	2.489	2.841	2.757	2.781
	C_2	3.914	3.898	3.854	3.834	3.782	4.476	4.322	4.175
	C_3	4.628	4.596	4.526	4.492	4.416	5.913	5.691	5.154
	C_4	5.305	5.250	5.179	5.133	5.051	7.199	6.926	5.762
Stratocumulus top	C_1	2.596	2.586	2.580	2.579	2.575	2.907	2.841	2.824
	C_2	3.978	3.963	3.950	3.941	3.934	4.723	4.584	4.370
	C_3	4.740	4.715	4.705	4.690	4.693	6.441	6.213	5.635
	C_4	5.413	5.395	5.362	5.362	5.372	8.074	7.746	6.670
Nimbostratus base	C_1	2.575	2.559	2.535	2.531	2.524	2.867	2.796	2.797
	C_2	3.935	3.911	3.875	3.864	3.842	4.575	4.436	4.244
	C_3	4.663	4.621	4.563	4.550	4.522	6.126	5.910	5.319
	C_4	5.337	5.294	5.248	5.205	5.169	7.555	7.258	6.066
Nimbostratus top	C_1	2.597	2.592	2.584	2.585	2.588	2.910	2.844	2.838
	C_2	3.981	3.969	3.958	3.954	3.955	4.735	4.598	4.424
	C_3	4.750	4.727	4.709	4.712	4.733	6.467	6.245	5.760
	C_4	5.419	5.399	5.379	5.384	5.420	8.123	7.806	6.892
Stratus base	C_1	2.555	2.537	2.520	2.507	2.476	2.829	2.731	2.770
	C_2	3.901	3.869	3.839	3.813	3.756	4.428	4.247	4.133
	C_3	4.602	4.543	4.499	4.456	4.371	5.805	5.547	5.058
	C_4	5.268	5.206	5.149	5.096	4.999	7.016	6.713	5.596
Stratus top	C_1	2.586	2.568	2.562	2.554	2.546	2.886	2.815	2.808
	C_2	3.954	3.932	3.916	3.900	3.881	4.643	4.499	4.294
	C_3	4.696	4.661	4.634	4.612	4.594	6.269	6.039	5.445
	C_4	5.367	5.339	5.300	5.275	5.253	7.789	7.466	6.310
C.5	C_1	2.586	2.565	2.553	2.545	2.533	2.876	2.803	2.802
	C_2	3.954	3.920	3.901	3.883	3.858	4.607	4.461	4.267
	C_3	4.696	4.637	4.609	4.584	4.555	6.191	5.962	5.379
	C_4	5.360	4.305	5.269	5.243	5.209	7.661	7.346	6.186
C.6	C_1	2.637	2.630	2.627	2.644	2.676	2.929	2.867	2.928
	C_2	4.070	4.049	4.037	4.059	4.114	4.813	4.692	4.782
	C_3	4.923	4.875	4.852	4.911	5.049	6.652	6.465	6.558
	C_4	5.634	5.563	5.521	5.614	5.832	8.460	8.205	8.269
Best	C_1	2.511	2.479	2.459	2.437	2.409	2.763	2.569	2.605
	C_2	3.809	3.753	3.708	3.657	3.577	4.167	3.730	3.609
	C_3	4.440	4.344	4.268	4.184	4.050	5.218	4.522	4.047
	C_4	5.062	4.961	4.852	4.734	4.520	5.994	5.129	4.078

largest liquid water contents occur for drop size distributions characteristic of cloud top while the distributions representing cloud bases are the more sharply peaked. Curve F representing the drop size distribution found at the top of nimbostratus clouds has the smallest peak values, but has the widest distribution. Therefore, it would appear that narrow drop size distributions found near cloud bases are characterized by smaller liquid water contents and smaller attenuation parameters (Table 2.2).

The phase function represents the proportion of total energy scattered in a particular direction. The value $f = C_4/9.0$ is a representation of the amount of energy contained in the forward-scattering peak (Wiscombe, 1977). Plots of various phase functions are given in Zdunkowski *et al.* (1967). In the delta-Eddington and delta-spherical harmonics methods, f represents the proportion of total scattered energy which is assumed transmitted without scattering during interaction with a water drop; conversely, $1 - f$ is the proportion scattered during this interaction. From Table 2.3 values of f typically vary from about 0.44 to 0.67. Similar values to those given in Table 2.3 are found in the 0.3–0.8 μm wavelength region. As would be expected, increased scattering leads to increased cloud albedo. The topic of drop size distribution effects upon the radiation field is developed more fully in a later section.

The C.5 drop distribution function is representative of nimbostratus clouds and is based upon the measurements of Okita (1961). As pointed out by Deirmendjian (1975), the C.5 distribution coincides almost exactly with measured distributions for cumulus congestus and nimbostratus clouds (Hansen, 1971). The C.6 distribution is representative of large droplet spectra often found in precipitating clouds.

2.2 *Method of calculation*

The present investigation utilizes the spherical harmonic technique developed by Zdunkowski and Korb (1974) and applied to absorption in clouds by Welch *et al.* (1976). Each absorbing interval is divided into N sub-intervals each of which is characterized by one b_i, K_i value [Eq. (2.1)]. The optical path length for each sub-interval is given by

$$\tau_i = \int_{\zeta_1}^{\zeta_2} [\bar{\beta}_e(\zeta') + K_i\,\rho(\zeta')]\,d\zeta', \tag{2.6}$$

where ζ_1 and ζ_2 are the integration limits at the cloud layer boundaries, $\bar{\beta}_e(\zeta')$ is the droplet extinction coefficient averaged over each spectral interval, and $\rho(\zeta')$ is the pressure-corrected water vapor density. The corresponding absorption quantity k is defined as

$$k_i(\zeta') = \frac{\bar{\beta}_a(\zeta') + K_i\,\rho(\zeta')}{\bar{\beta}_e(\zeta') + K_i\,\rho(\zeta')} = 1 - \tilde{\omega}_0, \tag{2.7}$$

where $\bar{\beta}_a(\zeta')$ is the droplet absorption coefficient averaged over each spectral interval and $\tilde{\omega}_0$ is the single-scattering albedo. The cloud model assumes that droplet concentration and size distribution are constant throughout each layer and that values of τ_i and k_i are held constant throughout each of the horizontally homogeneous cloud layers.

2.2.1 Delta function method

The phase function is approximated by a Dirac delta function for the forward-scattering peak (Wiscombe, 1977), i.e.,

$$P(\theta) = 2f\,\delta(1 - \cos\theta) + (1 - f)\sum_{\ell=0}^{N} C'_\ell P_\ell(\cos\theta), \tag{2.8}$$

where f represents the percentage of energy in the forward-scattering peak. Applying this transformation, values of τ, k and P_ℓ in the radiative transfer solution presented by Zdunkowski and Korb (1974) are now replaced with the values τ', k' and C'_ℓ, respectively, where

$$\tau' = [1 - (1 - k)f]\,\tau, \tag{2.9}$$

$$k' = \frac{k}{1 - (1 - k)f}, \tag{2.10}$$

$$C'_\ell = \frac{(2\ell + 1)(C_\ell/(2\ell + 1) - f)}{1 - f}. \tag{2.11}$$

2.2.2 Comparison with adding method

The delta-spherical harmonic technique is unstable for large optical depths ($\tau > 35$). Therefore, for large optical depths, the adding method using the diamond initialization (Wiscombe, 1976) has been utilized without the delta-function approach. Intercomparisons of fluxes and heating rates between the delta-spherical harmonics (δ) and adding (a) methods are given in Table 2.4. Values of upward diffuse flux ($F\uparrow$), total downward flux ($F\downarrow$), net flux (F_N) and heating rates ($\partial\theta/\partial t$) are given for the two methods of calculation. Thicknesses are measured downward from cloud top ($z = 0$). A 1000 m thick cloud with base height of 1000 m is assumed: with liquid water content (w_L) of 0.1 g m^{-3}, solar zenith angle of $\theta = 0°$ and the Best droplet size distribution. Table 2.5 shows average temperature (T) and specific humidity (Q) values observed during the GATE (Phase III) field measurements and used in the calculations reported in Table 2.4. Saturated water vapor concentrations were assumed within the cloud, and values were taken from the *Smithsonian Meteorological Tables* (1966). Table 2.4 shows the flux differences calculated with these two methods are slight. The adding method was computationally an order of magnitude slower than the delta-spherical harmonics technique for this particular case. However, it should be noted that the computational speed of the adding method varies significantly with cloud thickness and the number of internal levels at which fluxes are determined. The adding and delta-spherical harmonics technique are comparable in computation time if only bulk cloud properties are desired. However, if values at internal levels are required (in order to determine heating rates within the cloud, for instance) the total computer time is approximately the product of the time to compute radiative properties of the entire cloud and the number of layers into which the cloud has been divided. For the delta-spherical harmonics technique, values at internal levels are calculated at essentially little extra expense. Since the results are for all practical purposes equivalent for these two radiative transfer methods, the delta-spherical harmonic method is used in subsequent calculations for $\tau < 35$.

Table 2.4. Comparison between delta-spherical harmonics and adding methods for GATE profiles in Table 2.5. Cloud base height is 1000 m, thickness is 1000 m, surface albedo (A_s) is 0.04 (sea surface) and solar zenith angle (θ) is 0° using the Best (1951) droplet size distribution function for liquid water content (w_L) of 0.1 g m^{-3}. $F\uparrow$ refers to upward diffuse fluxes, $F\downarrow$ to total downward fluxes, and ($\partial\theta/\partial t$) to heating rates. Fluxes are given in W m^{-2} and heating rates in °C h.$^{-1}$ z is the depth (m) into the cloud measured from cloud top. Heating rates are given for the layers extending from cloud top to the level z.

z	$F\uparrow_\delta$	$F\uparrow_a$	$F\downarrow_\delta$	$F\downarrow_a$	$F_{N\delta}$	F_{Na}	$\frac{\partial\theta}{\partial t_\delta}$	$\frac{\partial\theta}{\partial t_a}$
0	778	781	1066	1066	288	285		
100	862	864	1124	1122	262	258	0.97	1.0
500	507	495	716	701	209	206	0.57	0.57
1000	7.7	7.7	202	198	194	190	0.34	0.34

TABLE 2.5. Pressure (P), temperature (T) and specific humidity (Q) from GATE observations averaged over Phase III.

P (mb)	T (°C)	Q (g kg^{-1})	P (mb)	T (°C)	Q (g kg^{-1})
100.0	−76.10	0.00	600.0	0.50	5.20
150.0	−69.00	0.00	650.0	4.50	6.20
200.0	−55.40	0.00	700.0	8.20	7.50
250.0	−42.90	0.00	750.0	11.40	8.80
300.0	−32.60	0.40	800.0	14.20	10.20
350.0	−24.30	0.80	850.0	16.60	11.90
400.0	−17.40	1.20	900.0	19.20	13.80
450.0	−11.70	1.90	950.0	22.00	15.90
500.0	−6.90	2.90	1000.0	25.60	18.00
550.0	−2.70	4.00	1012.00	27.00	17.70

2.2.3 SENSITIVITY OF BULK RADIATIVE PROPERTIES TO VARIATIONS IN PHASE FUNCTION EXPANSION COEFFICIENTS

Using his delta-M method, Wiscombe (1977) discusses cloud albedo error for both Mie and Henyey-Greenstein phase function approximations. He further shows that the delta-M method leads to remarkably accurate computed fluxes at very low orders of angular approximation no matter how great the asymmetry in the phase function. The question still remains as to how sensitive the cloud reflectance (R), absorptance (A) and heating rates $(\partial\theta/\partial t)$ are to variations in the phase function expansion coefficients. Table 2.6 considers a number of these cases as a function of wavelength.

TABLE 2.6. Reflectance R and absorptance A (percent) and cloud-averaged heating rates [$\partial\theta/\partial t$ (°C h^{-1})] as a function of wavelength region (μm) for a cloud of thickness 500 m with base height of 2 km at solar zenith angle of 0°. Values of the phase function expansion coefficients (C_1, C_2, C_3) and $f = C_4/9$ have been scaled as shown.

Case number and identification			← Wavelength region (μm) →									Average cloud heating rate (°C h^{-1})
			<0.8	0.95	1.15	1.4	1.8	2.8	3.3	6.3	Total	
1	Standard case	R	64.8	57.9	54.7	46.6	42.6	2.2	1.5	0.8	59.3	0.78
		A	0.001	8.0	16.3	33.3	43.1	97.2	98.5	99.1	9.0	
2	f*1.1	R	64.7	57.9	54.6	46.4	42.4	1.6	1.3	0.6	59.2	0.79
		A	0.002	8.0	16.4	33.4	43.2	97.8	98.7	99.3	9.1	
3	f*1.25	R	64.6	57.7	54.4	46.2	42.1	0.6	0.9	0.4	59.0	0.79
		A	0.002	8.1	16.5	33.6	43.5	98.8	99.1	99.6	9.1	
4	f*1.5	R	64.5	57.4	54.0	45.6	41.6	0.5	0.3	1.4	58.8	0.80
		A	0.004	8.2	16.8	34.1	44.0	96.9	99.6	98.6	9.2	
5	f*0.9	R	64.9	58.0	54.8	46.8	42.7	2.8	1.7	1.0	59.4	0.78
		A	0.001	7.9	16.3	33.1	42.9	96.6	98.3	98.9	9.0	
6	f*0.75	R	65.0	58.1	55.0	47.0	43.0	3.6	2.1	1.3	59.5	0.77
		A	0.001	7.9	16.2	33.0	42.7	95.8	97.9	98.7	8.9	
7	f*0.5	R	65.1	58.3	55.2	47.3	43.3	4.8	2.6	1.8	59.7	0.77
		A	0.001	7.8	16.1	32.7	42.4	94.6	97.4	98.2	8.9	
8	C_1*1.1	R	51.0	40.8	39.2	33.0	30.4	9.3	1.4	2.4	45.0	0.84
		A	0.001	8.6	17.4	35.6	46.5	90.7	98.6	97.6	9.7	
9	C_2*1.1	R	64.5	57.8	54.7	47.0	43.1	5.0	2.4	1.7	59.2	0.76
		A	0.001	7.7	15.9	32.4	42.2	94.3	97.6	98.3	8.8	
10	C_3*1.1	R	65.1	58.4	55.2	47.2	43.2	3.5	1.9	1.2	59.6	0.77
		A	0.001	7.8	16.1	32.7	42.5	95.9	98.1	98.8	8.9	
11	C_1*0.75	R	79.3	74.6	70.5	61.6	57.1	19.1	7.7	7.6	74.3	0.68
		A	0.001	6.9	14.3	29.0	36.8	80.7	92.2	92.4	7.9	
12	C_2*0.75	R	65.3	58.3	54.9	46.3	42.0	1.8	0.3	1.0	59.6	0.81
		A	0.001	8.2	16.8	34.2	44.2	98.2	99.5	99.0	9.3	
13	C_3*0.75	R	64.4	57.2	53.9	45.5	41.4	0.2	0.6	0.04	58.6	0.80
		A	0.001	8.2	16.8	34.2	44.2	99.7	99.4	99.9	9.3	

Case 1 presents a 500 m thick cloud with base height of 2 km and solar zenith angle of 0° using the Best drop size distribution for a liquid water content of 0.1 g m^{-3}. Cloud total reflectance is 59.3% and cloud absorptance is 9%. Cases 2–4 consider the situations in which f, the percentage of energy in the forward-scattering peak, is arbitrarily increased by 10%, 25% or 50%, respectively. Bulk cloud properties R, A and $\partial\theta/\partial t$ are surprisingly insensitive to the value of f, with the largest variation at larger wavelengths. Increasing f leads to less energy for scattering with somewhat lower values of reflectance. In Cases 5–7 f is decreased to 90%, 75% or 50% of its original value, respectively. This situation leads to increased scattering since there is less energy in the forward-scattering peak. As expected cloud reflectance values have increased, but these changes are not significant. Cases 8–10 consider each of the phase function expansion coefficients (C_1, C_2, C_3) separately increased by 10%. Case 8 shows that the results are highly sensitive to the value of C_1, or to the symmetry factor $g = \langle \cos\theta \rangle = C_1/3$. Significantly lower values of reflection occur at all wavelengths for this case along with increased absorption. However, variations in C_2 and C_3 (cases 9 and 10) once again have little effect upon the bulk cloud properties. Finally, cases 11–13 consider these same expansion coefficients arbitrarily decreased to 75% of their former values. Case 11 shows the opposite effect noted in case 8; decreases in the asymmetry factor g or in C_1 lead to increased scattering (increased reflectance) and decreased absorptance. Once again variations in the C_2 and C_3 coefficients, even by 25%, have little effect on total cloud properties. These results support Wiscombe's (1977) statement that the reason for the success of the delta-Eddington approach (as well as for the delta-spherical harmonics method) is that fluxes are most sensitive to the lowest order moments; therefore, the selection of $f = \chi_{2m}$, used in the delta-M method, appears to be the most consistent choice and far better than the choices of truncated phase functions used in the past. The sensitivity of the bulk cloud radiative properties to the C_1 (or g) expansion coefficient and their relative sensitivity to the higher order coefficients (even the C_2 term) helps to explain the surprising success of the Henyey-Greenstein approach. In this formalism the phase function expansion coefficients take the form

$$C_\ell = (2\ell + 1)g^\ell. \tag{2.12}$$

Note, however, that $g = C_1/3$ and that the first term reduces once again to C_1. Since the exact value of the higher order terms does not affect radiative fluxes significantly, the Henyey-Greenstein approximation is a reasonable representation of the scattering phase function. In terms of Wiscombe's delta-M method, g is replaced by g'. For the four-term delta-spherical harmonics approach

$$f = g^4, \tag{2.13}$$

$$g' = \frac{g - g^4}{1 - g^4}, \tag{2.14}$$

as shown by Joseph *et al.* (1976).

2.3 Applications to the GATE region

2.3.1 Cloud heating rates

Fig. 2.2 shows cloud heating rates (°C h^{-1}) as a function of cloud top height (z_T). Attenuation parameters and phase functions (Tables 2.2 and 2.3) corresponding to the Best droplet size distribution for liquid water contents of 0.1 and 0.2 g m^{-3} have been utilized. Cloud heating rates are shown as a function of cloud thickness [1000 m (Figs. 2.2A,B), 500 m (Figs. 2.2C,D) and 200 m (Figs. 2.2E,F)] and solar zenith angle (0°, 30°, 60°) using the GATE Phase III water vapor profile and an assumed surface albedo of $A_s = 0.04$ (sea surface). Since cloud heating rates are highly variable, with the largest values within 100 m of the cloud top [Table 2.4 and Welch *et al.* (1976)], average cloud heating rates are given for the upper 50 m (Figs. 2.2 A,C,E) as well as for the total cloud thickness (Figs. 2.2 B,D,F).

Fig. 2.2 shows that heating rates vary slowly between zenith angles of 0° and 30° and more rapidly between zenith angles of 30° and 60° in all cases. In Figs. 2.2 A,C,E we show that average heating rates in the upper 50 m of clouds at a given cloud top height are highly insensitive to cloud thickness. Solar radiation is rapidly attenuated by scattering and absorption as it penetrates down into the cloud. Only a small fraction of the total radiation penetrating the cloud top is transmitted through the cloud base (Table 2.4) for a 1000 m thick cloud. Therefore, average cloud heating rates decrease rapidly with increasing cloud thickness. It should be reiterated at this point that the above results apply for horizontally infinite and homogeneous clouds without vertical structure. The effects of horizontal and vertical structure are considered in later sections.

Welch *et al.* (1976) showed that droplet absorption in the water vapor bands is a significant contributor to large cloud heating rates. An unanswered question remains: What is the relative contribution of water vapor to cloud absorption and heating rates? Calculations for the same cloud conditions used in Table 2.4, but neglecting droplet absorption and scattering, provide heating rates of 0.23°C h^{-1} in the upper 50 m of the cloud with a cloud average heating rate of 0.16°C h^{-1}. For the cloud model assumed in Table

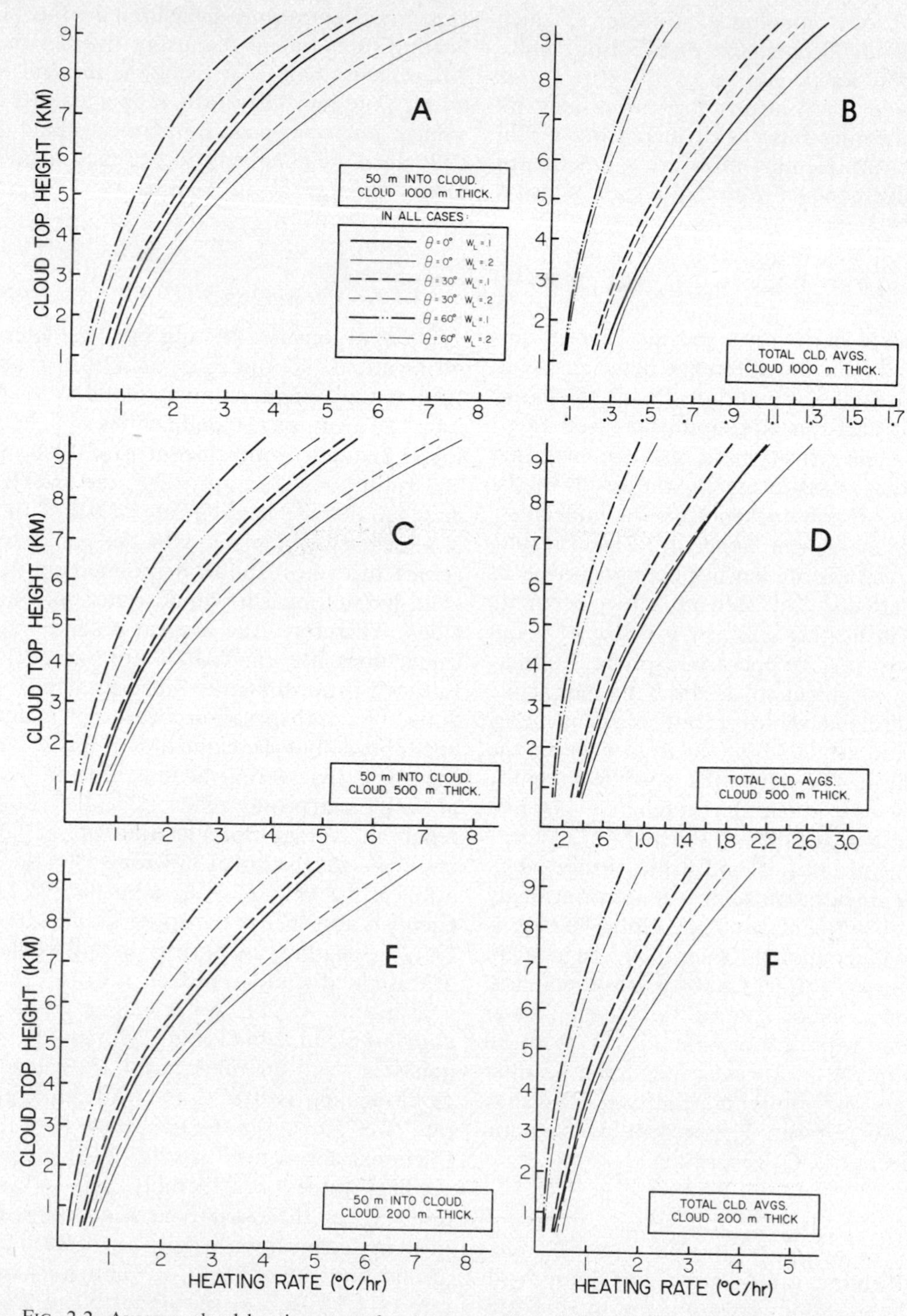

FIG. 2.2. Average cloud heating rates, for the upper 50 m and total cloud for clouds of varying thickness (200, 500 and 1000 m) at varying solar zenith angles (θ = 0°, 30° and 60°); liquid water content is 0.1 or 0.2 g m^{-3}.

2.4, the average heating rate in the upper 50 m is 1.03°C h^{-1} with a cloud average rate of 0.34°C h^{-1}. Water vapor contributes approximately 20% of cloud solar heating near the cloud top, but up to 50% of cloud heating averaged over this 1000 m thick cloud model. Therefore, we conclude that droplet absorption is primarily responsible for absorption of solar radiation in thin clouds and roughly of equal importance to water vapor in thick clouds. It appears that the water vapor contribution with respect to droplets increases as the radiation field becomes increasingly diffuse. The above conclusions apply only to droplet size distributions which are monomodal. Chapter 3, which considers the bimodal droplet size distribution, modifies this conclusion drastically.

We now consider the reasons for the striking increase of solar heating rates with increasing height as shown in Fig. 2.2. Heating rates are a function of radiative flux divergence ($\partial F_N/\partial z$) and air density, i.e.,

$$\frac{\partial \theta}{\partial t} = \frac{1}{\rho\, c_p} \frac{\partial F_N}{\partial z}, \tag{2.15}$$

where ρ is the air density, c_p is specific heat at constant pressure and F_N is the difference between downward and upward fluxes, or net flux. Eq. (2.15) shows that for constant energy absorption, increased cloud height (decreasing air density) will result in increased cloud heating rates. This is merely the result of the same energy being distributed among fewer molecules, leading to more energy per molecule. These results consider only a static situation neglecting latent heat and dynamical effects. This density effect accounts for an increase in heating rates by a factor of about 3 between the surface and upper troposphere. Furthermore since there is significantly less water vapor above the higher altitude clouds there is greater incident energy upon the cloud top available for absorption within the cloud. Using the GATE Phase III specific humidity values, the total solar radiation reaching cloud top at a zenith angle of $\theta = 0°$ is 883, 1182 and 1277 W m^{-2} at cloud top heights of 1, 5 and 8 km respectively. Within the water vapor bands solar radiation is strongly absorbed in the band centers. Since cloud heating is so strongly dependent upon the solar radiation incident on droplets in the near-infrared, absorption of incoming radiation by water vapor can be significant. Water droplet and water vapor absorption within the cloud dominates that of CO_2. However, depletion of solar energy reaching cloud top within the CO_2 band may be significant in some cases. The present investigation neglects the effects of CO_2 absorption.

2.3.2 Variations in water vapor profile

Since cloud heating rates vary significantly with cloud height, the question arises whether or not seasonal variations in the water vapor profile may lead to significant variations in cloud bulk radiative properties. Water vapor profiles taken from McClatchey *et al.* (1971) for midlatitude winter, midlatitude summer and tropical atmospheres were substituted for the GATE water vapor profile. The 1000 m thick cloud model with cloud top at 2 km given in Table 2.4 was used. Cloud average heating rates varied less than 10% for all three cases. However, due to the large water vapor concentrations measured in the GATE area, heating rates calculated using the GATE data were 20% smaller than those using the tropical values given by McClatchey. Therefore, it appears that clouds with similar microphysical properties, cloud thicknesses and cloud heights may experience significantly different average cloud heating rates in various geographical locations.

2.3.3 Comparisons with observations

We now consider heating rates as determined from observations. Reynolds *et al.* (1975) reported that a 1200 m thick stratocumulus cloud with cloud top at 3 km had an average cloud heating of 0.36°C h^{-1}. From Fig. 2.2B with cloud thickness of 1000 m, cloud top at 3 km, $w_L = 0.2$ and $\theta = 30°$, the GATE model predicts an average heating rate of 0.33°C h^{-1}, which is in excellent agreement with the results of Reynolds *et al.* However, it should be pointed out that Reynolds *et al.* did not measure liquid water content within this cloud. Therefore, the present results only show that calculations are capable of producing cloud heating rates which are similar to those obtained from observations. While observations were not available for the upper portion of the cloud alone, Fig. 2.2A shows that the calculated average heating rates in the upper 50 m of such a cloud may reach 1.7°C h^{-1}. Reynolds *et al.* report an average cloud heating rate of 0.32°C h^{-1} and fractional absorption of 36% for a 5 km thick cloud with a top at 5.2 km. If we assume that 95% of the total energy is absorbed in the upper 1 km of the cloud (Table 2.4), the heating rate corresponding to the upper 1 km of this cloud could approach 1.5–1.6°C h^{-1}.

Using the GATE water vapor profile and stratocumulus cloud data (Tables 2.2 and 2.3) with $w_L = 0.3$ g m^{-3} for this 5 km thick cloud, the calculated average cloud heating is 0.16°C h^{-1} with an average heating rate of 0.75°C h^{-1} in the upper 500 m of the cloud and fractional absorption of 17%. In this case the model predicts values considerably smaller than observations. Using these same stratocumulus data, a cloud of 1 km thickness with cloud top at 2 km gives an average heating rate of 0.33°C h^{-1} with a maximum heating rate of 1.1°C h^{-1} in the upper 50 m and total cloud absorption of 9%. Therefore, increasing the cloud thickness from 1 to 5 km has only increased the absorption from 9% to 17%. Reynolds *et al.* report absorption values of up to 52%, or half the total solar energy incident upon the cloud top. Twomey (1976) and Stephens (1978) show that even for very thick clouds, the maximum absorption is only 20%. However, their results are based upon the use of a monomodal droplet size distribution.

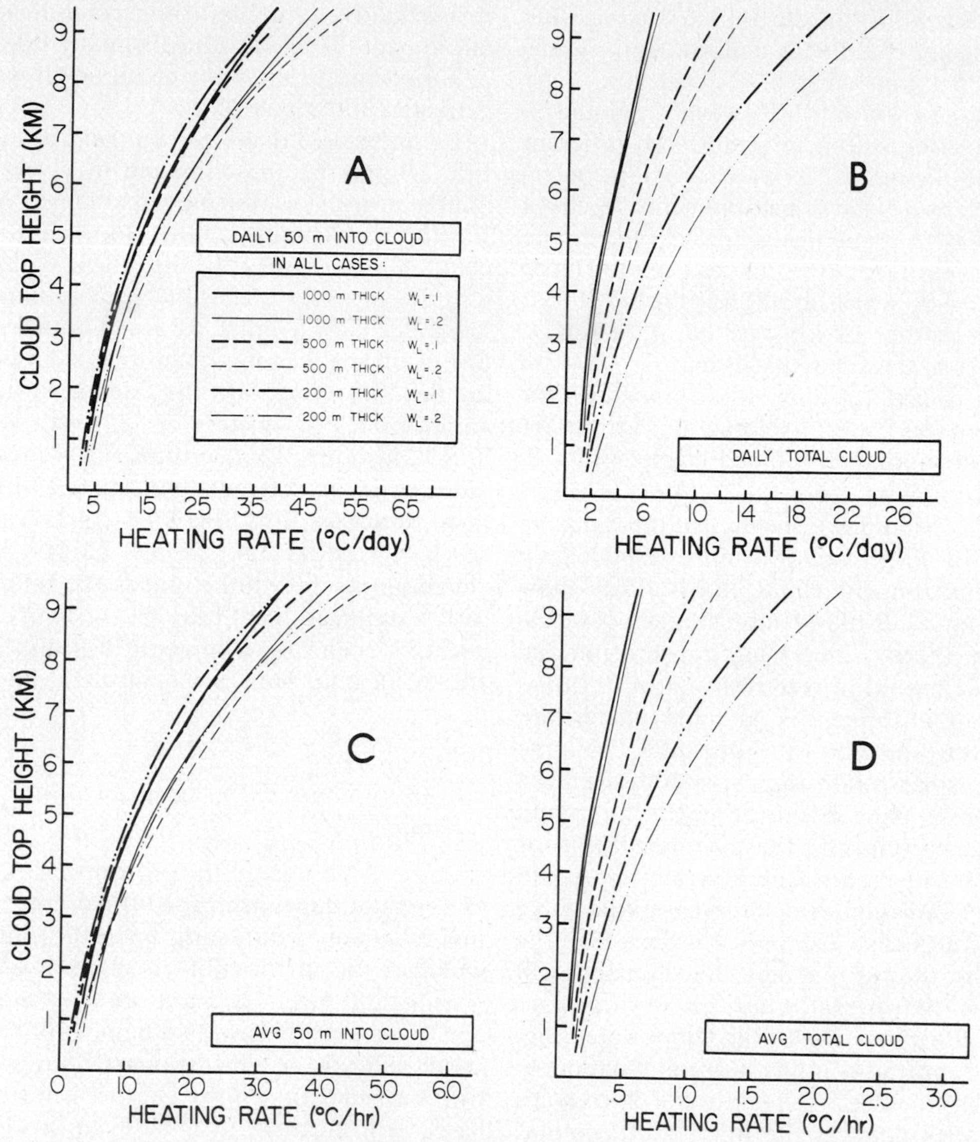

FIG. 2.3. Integrated daily average cloud heating rates versus cloud top height for clouds of varying thickness (200, 500 and 1000 m) with liquid water contents of 0.1 or 0.2 g m^{-3}.

2.3.4 Integrated daily heating rates

Fig. 2.3 shows tropical integrated daily solar heating rates as a function of cloud top height for clouds of thickness 200, 500 and 1000 m with liquid water contents of $w_L = 0.1$ and 0.2 g m^{-3}. These tropical values are nearly independent of latitude and season. Fig. 2.3A shows the integrated total daily heating rate averaged over the upper 50 m, while Fig. 2.3B shows the corresponding values averaged over the entire cloud. Figs. 2.3C and 2.3D give average daily heating rates for the upper 50 m and total cloud thickness. Calculations have been performed in 20 min time steps for solar zenith angles $< 84°$ (i.e. $\mu_0 = \cos\theta_0 \geq 0.1$). The values reported in Figs. 2.3A and 2.3C are very large. Clouds subject to such large heating rates at cloud top would be expected to rapidly "burn off" in the absence of a compensating energy sink from other processes. The average cloud heating rates, however, do not seem to be excessive, even for high clouds in light of Reynolds *et al.* (1975) report of an average heating rate of 7.76°C day^{-1} for a 3 km thick cirrus cloud with top at 11 km.

2.3.5 Discussion of results

Ice equivalent liquid water contents in cirrus clouds generally increase with optical depth into the cloud. Griffith *et al.* (1980) have shown ice equivalent water contents in cirrus clouds varying from 0.04 g m^{-3} at the cloud top to 0.12 g m^{-3} at the cloud base. Considering a high-altitude water cloud with uniform liquid water content of w_L = 0.1 g m^{-3}, Fig. 2.3B gives an average daily heating rate of 7.3°C day^{-1}. However, it should be pointed out that ice crystals have significantly different optical properties compared to water droplets. Interestingly, Griffith *et al.* show measurements in cirrus clouds with radiative properties similar to water. The radiative properties of ice clouds will be explored more fully in Chapter 4 in which it will be concluded that the radiative properties of ice may be quite similar to those of water in the solar spectrum.

The question of large heating rates computed near the cloud top requires further exploration. The present calculations assume a uniform cloud layer, both vertically and horizontally. McKee and Cox (1974, 1976) and Davis *et al.* (1979a) have shown that the radiative characteristics of finite clouds can be significantly different from horizontally infinite clouds. They show that significant amounts of radiation escape out of the sides of clouds, thereby increasing transmission and decreasing reflection. However, Davis *et al.* (1979a) show that cloud absorptances in finite clouds are smaller than absorptances in horizontally infinite clouds at small solar zenith angles, and absorptance values are larger in finite clouds at large solar zenith angles. While the present model assumes homogeneous parameters, real clouds are highly variable in liquid water content. Radiation may be transmitted large distances if channels of smaller optical density material exist within the overall average cloud conditions. Platt (1976) has shown that while size distributions remain similar at a given height, the liquid water content may vary appreciably in the horizontal direction. Squires (1958) has reported the presence of cloud holes devoid of droplets deep in the interior of cumulus clouds and suggested that they are produced by the entrainment of pockets of dry air into the tops of cloud towers. Mason and Jonas (1974) have modeled this situation by assuming that the process starts with the appearance of protuberances of the order of 100 m in diameter on the caps of emerging cloud towers which trap pockets of relatively dry environmental air from above the cloud. These then sink into the cloud as a result of evaporative cooling. This model, however, shows that deeply penetrating downdrafts and holes are expected mainly in inactive or decaying clouds. In such cases, reflection would decrease resulting in smaller cloud albedos, transmission would increase through the cloud, and cloud heating would be distributed more uniformly throughout the cloud. Nevertheless, strong solar heating must still occur in the upper cloud deck in excess of average cloud solar heating rates. The average cloud heating rates from the present investigation are probably reasonably representative of actual cloud conditions while those of the upper 50 m are unrealistically large. The effect of horizontal variations in cloud liquid water contents is treated in Chapter 6.

The integrated daily average heating rates shown in Fig. 2.3 may be approximated in a very simple way for the tropical cloud model assumed in Table 2.4. For the horizontally infinite cloud case, a single calculation with a solar zenith angle of approximately θ = 50° provides a reasonable approximation of the integrated daily average values of both radiative fluxes and cloud heating rates. The solar zenith angle providing the "closest fit" to the integrated daily averages varied from θ = 46° to θ = 53° under various conditions. However, a value of θ = 50° consistently provided estimates of radiative fluxes and heating rates good to within 10%. Davis *et al.* (1979a) find similar results for horizontally infinite clouds. However, for small, horizontally finite clouds the solar zenith angle which provides the "best fit" with integrated daily averages is shown to be about θ = 40°. These results are, of course, both seasonally and latitudinally dependent.

2.4 *The effect of drop size distributions upon the radiation field*

Since cloud heating rates, transmission and reflection may be dependent upon the droplet size distribution functions, particularly upon the mode radius and width of the distribution, a number of droplet size distribution functions have been studied. Tampieri and Tomasi (1976) have fit a number of cloud measurements using the modified gamma distribution proposed by Deirmendjian (1969). The present investigation is based upon measurements by Tverskoi (1965) in stratocumulus, nimbostratus and stratus clouds (Tables 2.1–2.3) at both cloud base and cloud top. While many other distribution functions could have been chosen, it is believed that these represent a reasonable variety of conditions.

2.4.1 Cloud heating rates

Fig. 2.4 shows heating rates (°C h^{-1}) as a function of cloud top height for a 1000 m thick cloud. Liquid water content is scaled to w_L = 0.1 g m^{-3} in order to more easily compare results using these various droplet size distributions. Nevertheless, it should be noted that scaling of attenuation parameters according to liquid water content is not generally valid. This is due to the fact that with increasing liquid water content, there are increasing numbers of larger droplets whose scattering and absorption properties are significantly different from those of smaller droplets (Paltridge, 1974b).

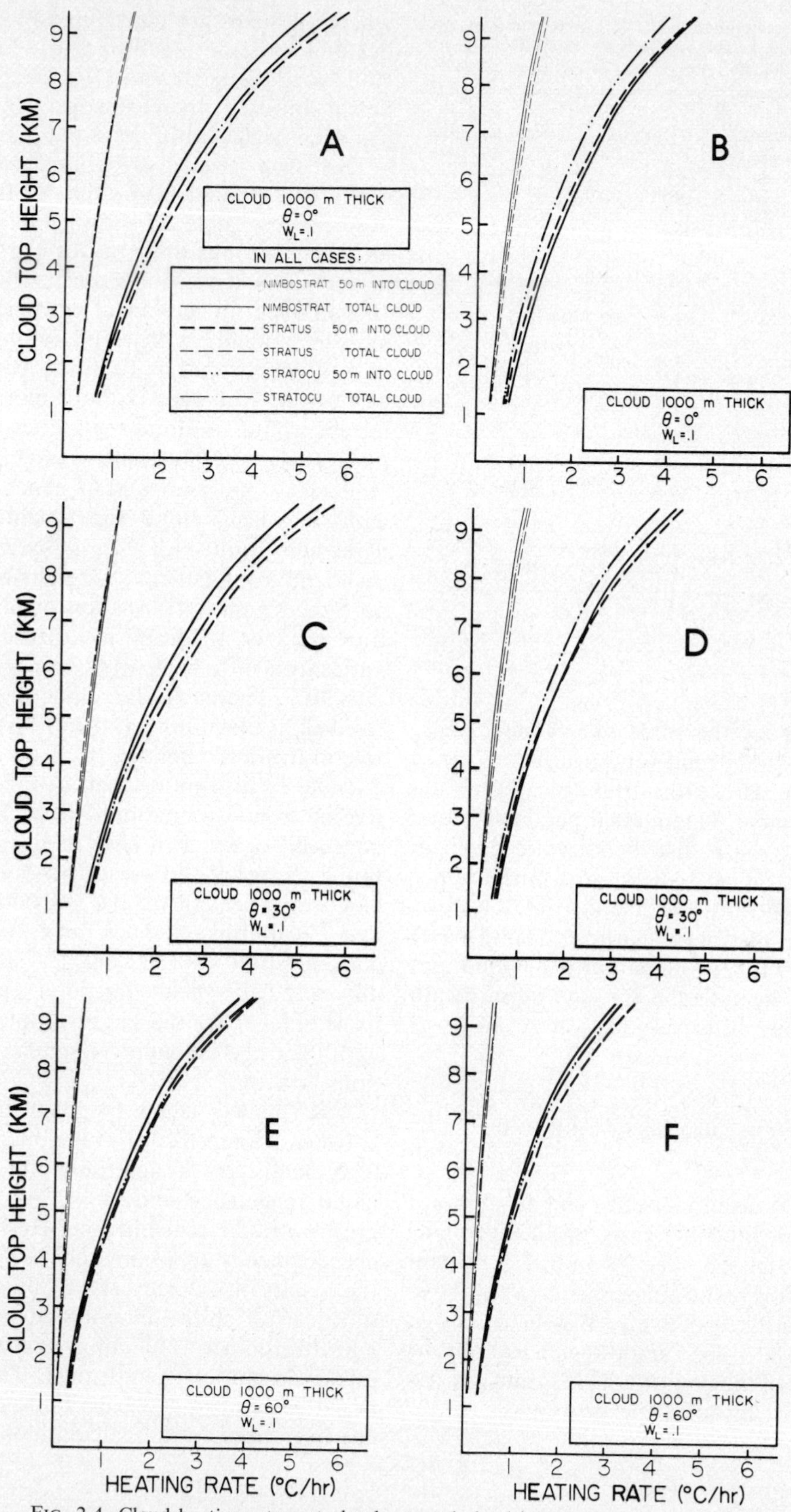

FIG. 2.4. Cloud heating rates, at cloud top and cloud base versus cloud top height for various types of clouds all 1000 m thick. Values are plotted for three solar zenith angles ($\theta = 0°$, 30° and 60°); liquid water content is 0.1 g m^{-3} in all cases. Cases Ⓐ, Ⓒ and Ⓔ were calculated using a drop size distribution characteristic of cloud base (i.e., small-particles); in Ⓑ, Ⓓ and Ⓕ it was characteristic of cloud top (i.e., large-particles).

TABLE 2.7. Percent reflectance (R), transmittance (T) and absorptance (A) of a 1000 m thick cloud at selected cloud top heights for six drop size distributions at a solar zenith angle of 0°.

Cloud top height (km)		Nimbostratus Top	Nimbostratus Base	Stratus Top	Stratus Base	Stratocumulus Top	Stratocumulus Base
1.5	R	80.1	70.2	74.1	62.0	79.0	65.0
	T	8.7	20.5	16.0	29.3	10.0	26.0
	A	11.2	9.3	9.9	8.7	11.0	9.0
4	R	78.0	68.7	72.4	61.0	77.0	63.9
	T	8.2	19.6	15.3	28.3	9.5	25.0
	A	13.8	11.7	12.3	10.7	13.5	11.1
6	R	76.1	67.9	71.0	60.2	75.2	62.9
	T	7.9	19.0	14.7	27.5	9.1	24.3
	A	16.0	13.6	14.3	12.3	15.7	12.8
8	R	75.0	67.1	70.4	60.2	74.3	62.8
	T	7.7	18.7	14.5	27.2	8.9	24.0
	A	17.3	14.2	15.2	12.6	16.8	13.2
10	R	73.6	66.3	69.4	59.7	73.0	62.2
	T	7.5	18.4	14.2	26.9	8.7	23.7
	A	18.9	15.3	16.4	13.4	18.3	14.1

Fig. 2.4 shows that for the same solar zenith angle, average heating rates are relatively invariant to cloud droplet distribution. However, heating rates in the upper 50 m are variable. Attenuation parameters and phase functions for these distributions are given in Tables 2.2 and 2.3. Fig. 2.4 shows that in the upper cloud region, heating rates for size distribution functions representative of smaller particles (cloud base) are larger than those for the larger particle (cloud top) distributions. At a height of 8 km and solar zenith angle of $\theta = 0°$, the difference may be as large as 0.5–1.0°C h^{-1}.

2.4.2 REFLECTANCE, TRANSMITTANCE AND ABSORPTANCE

Reflectance (R), transmittance (T) and absorptance (A) are normally defined in terms of the incoming radiative flux at cloud top. Let T_D and T_E represent the total optical depths of the atmospheric layers above and within the cloud, respectively. With $\mu_0 = \cos\theta$, where θ refers to the solar zenith angle, R T and A are defined in terms of the downward $(F\downarrow)$ and upward $(F\uparrow)$ fluxes and parallel solar energy (E_0) as

$$R = \frac{F\uparrow(\tau_E = 0)}{\mu_0 E_0 \exp(-T_D/\mu_0)}, \tag{2.16}$$

$$T = \frac{F\downarrow(\tau_E = T_E) + \mu_0 E_0 \exp[-(T_D + T_E)/\mu_0]}{\mu_0 E_0 \exp(-T_D/\mu_0)}, \tag{2.17}$$

$$A = \frac{\{\mu_0 E_0 \exp(-T_D/\mu_0) - F\uparrow(\tau_E = 0)\} - \{\mu_0 E_0 \exp[-(T_D + T_E)/\mu_0] + F\downarrow(\tau_E = T_E)\}}{\mu_0 E_0 \exp(-T_D/\mu_0)}, \tag{2.18}$$

where all fluxes are considered to be positive quantities and τ_E is the cloud optical depth measured from cloud top. Rayleigh scattering decreases rapidly with height and with increasing wavelength; therefore, it has been neglected in these expressions.

The drop size distribution strongly affects the scattered radiation field while weakly affecting cloud absorptance. Table 2.7 shows cloud reflectance, transmittance and absorptance for these same drop size distributions at various cloud top heights for a cloud 1000 m thick. In agreement with Fig. 2.4, absorptance increases rapidly with cloud top height, by as much as 50% for a cloud at 10 km as compared to a cloud at 1.5 km. Not only is there more energy available for absorption as cloud top increases with height, but more of the available energy is being absorbed. Both reflectance and transmittance decrease very slowly with increasing cloud top height. Furthermore, in agreement with Fig. 2.4, absorptance is relatively invariant to cloud drop size distribution. However, note the significant variation of both reflectance and transmittance to drop size distribution. Reflectance varies from 60 to 80%, while transmittance varies from 8 to 30%. Therefore, the cloud drop size distribution (as well as cloud optical thickness) plays a significant role in the determination of cloud albedo.

Table 2.7 also shows that the reflectance representative of drop size distributions at the tops of clouds is normally larger than that of drop size distributions found in the cloud bases. This is due to the fact that cloud tops generally have larger droplet sizes. However, scattering is much more sensitive to droplet concentration than to droplet sizes; therefore, the diffuse radiative field near cloud top would be expected to be smaller for the larger droplet size distributions with their correspondingly smaller droplet concentrations.

Absorptance values for 1000 m thick clouds vary between 9 and 11% for low clouds and 14 to 19% for high clouds. Increasing the cloud thickness increases cloud reflectance and cloud absorptance, while decreasing cloud transmittance. However, the maximum absorptance is approximately 20%, in agreement with the results of Twomey (1976) and Stephens (1978).

Table 2.8 shows cloud reflectance, transmittance and absorptance as functions of cloud top height, cloud thickness and solar zenith angle. The drop size distribution representing the base of a stratocumulus cloud has been used for all calculations presented in Table

2.8. Similar behavior has been verified for other drop size distributions.

Table 2.8 shows that with increasing solar zenith angles reflectance increases and both transmittance and absorptance decrease. However, an examination of Tables 2.7 and 2.8 reveals that cloud bulk radiative properties R, T and A, referenced to the energy at cloud top, are more sensitive to drop size distribution than to variations in solar zenith angle (in the range $\theta = 0°$ to $60°$).

For a solar zenith angle of 0°, decreasing the cloud thickness by a factor of 2 strongly decreases both reflectance (albedo) and absorptance while strongly increasing cloud transmittance. Comparing Tables 2.7 and 2.8 shows that changes in cloud reflectance due to variations in cloud droplet size distributions are roughly the same magnitude as changes in reflectance caused by varying cloud optical depth by a factor of 2. For a very thin cloud (100 m thick) both reflectance and absorptance become small. However, note that absorptance increases very rapidly with cloud top height for thin clouds. This is principally the result of more incident energy near the absorption line centers in the case of the higher clouds.

2.4.3 Reflectance, transmittance and absorptance defined in terms of incident solar flux

Tables 2.7 and 2.8 provide values of R, T and A defined according to the amount of energy reaching cloud top. However, it is often useful for satellite observations to define R, T and A in terms of the total solar energy incident upon the top of the atmosphere. Table 2.9 shows values of R', T' and A' defined in this way for the identical situation presented in Table 2.7.

Table 2.9 shows that in terms of total incident energy, reflectance increases with increasing cloud top height rather than decreasing with increasing cloud top height as implied by Table 2.7. In terms of absolute energy (i.e., remote sensing applications) reflectance/cloud brightness may increase by more than 10% from low clouds to high clouds of the same optical density. Once again variations in drop size distributions may lead to variations in reflected energy by 10–15%. Decreasing cloud thickness by a factor of 2 also decreases reflected energy by 12–15%. Therefore, for remote sensing applications, cloud optical depth, cloud top height, cloud drop size distributions and cloud geometry (Chapter 5) are all important contributors to cloud brightness measurements.

A cloud at 10 km absorbs about twice as much energy as an identical cloud would at 1.5 km. Assuming a decrease in air density by a factor of about 3 between 1.5 and 10 km then provides an increase by a factor of 6 in heating rates [Eq. (2.12)] for these two clouds at different heights. Therefore, the heating rates presented in Figs. 2.2–2.4 are quite reasonable. We conclude that for identical cloud conditions, average

Table 2.8. Percent reflectance (R), transmittance (T) and absorptance (A) at selected cloud top heights for the drop size distribution representing the base of a stratocumulus cloud. Values for a 1 km thick cloud are given for three solar zenith angles ($\theta = 0°$, 30° and 60°) along with values at solar zenith angle of 0° for cloud thicknesses of 500 and 100 m.

Cloud top height (km)		Cloud thickness 1000 m			500 m	100 m
		$\theta = 0°$	$\theta = 30°$	$\theta = 60°$	$\theta = 0°$	$\theta = 0°$
1.5	R	74.1	76.0	82.3	63.1	24.5
	T	16.0	15.0	11.4	29.4	73.5
	A	9.9	9.0	6.1	7.5	2.0
4.0	R	72.4	74.3	80.9	62.0	24.4
	T	15.3	14.3	10.8	28.4	72.6
	A	12.3	11.4	8.3	9.6	3.0
6.0	R	71.0	73.0	79.4	61.0	24.2
	T	14.7	13.6	10.3	27.6	71.7
	A	14.3	13.4	10.3	11.4	4.1
8.0	R	70.4	72.3	78.5	60.2	24.1
	T	14.5	13.3	10.0	27.0	71.2
	A	15.2	14.4	11.5	12.8	4.7
10.0	R	69.4	71.2	76.8	59.5	23.9
	T	14.2	13.0	9.7	26.6	70.2
	A	16.4	15.8	13.5	13.9	5.9

Table 2.9. Percent reflectance (R'), transmittance (T'), absorptance (A') and absorption by water vapor above the cloud (ATM), scaled in terms of total incident radiation at the top of the earth's atmosphere at selected cloud top heights for six drop size distributions. Cloud thickness is 1000 m, $\theta = 0°$.

Cloud top height (km)		Nimbostratus Top	Nimbostratus Base	Stratus Top	Stratus Base	Stratocumulus Top	Stratocumulus Base
1.5	R'	63.4	55.7	58.7	49.1	62.6	51.4
	T'	7.0	16.4	12.8	24.0	8.0	21.4
	A'	8.9	7.6	8.0	6.9	8.7	7.1
	ATM	20.7	20.3	20.5	20.0	20.7	20.1
4.0	R'	66.8	59.1	62.1	52.3	66.0	54.7
	T'	7.2	17.0	13.3	25.1	8.3	22.3
	A'	11.9	10.2	10.7	9.2	11.6	9.5
	ATM	14.1	13.7	13.9	13.4	14.1	13.5
6.0	R'	69.8	62.1	65.2	55.3	69.0	57.6
	T'	7.3	17.6	13.7	26.1	8.4	23.1
	A'	14.8	12.6	13.3	11.3	14.4	11.7
	ATM	8.1	7.7	7.8	7.3	8.2	7.6
8.0	R'	72.2	64.8	67.8	57.9	71.4	60.3
	T'	7.5	18.2	14.1	27.1	8.7	23.8
	A'	16.7	13.8	14.7	12.1	16.3	12.7
	ATM	3.6	3.2	3.4	2.9	3.6	3.2
10.0	R'	73.5	66.3	69.4	59.7	73.0	62.2
	T'	7.5	18.4	14.2	26.9	8.7	23.7
	A'	18.9	15.3	16.4	13.4	18.3	14.1
	ATM	0.0	0.0	0.0	0.0	0.0	0.0

heating rates in high clouds may exceed those in low clouds by a factor of at least 5. The magnitude of these variations appear to be nearly independent of cloud drop size distribution.

Total energy absorbed by water vapor above the cloud is given by ATM in Table 2.9. For clouds with a top height of 1.5 km, water vapor absorption above the cloud is approximately 20% of the total incoming solar radiation reaching the top of the earth's atmosphere. These results are for the GATE water vapor profile. For a cloud at 10 km water vapor absorption above the cloud may be neglected, as cloud top height is decreased, Table 2.9 shows that the loss of energy (by gaseous absorption) affects cloud absorptance and cloud reflectance roughly equally. On the way down to the top of the cloud radiation is strongly attenuated near the peaks of the water vapor absorption bands. Therefore, as the cloud top is lowered the absorption by water vapor within the cloud becomes progressively weaker. The reflected energy decreases since energy is absorbed both on the way down to the cloud top and on the way back out to the top of the atmosphere. The reflected energy is also diffuse with an average increased path length over the direct beam. In any case, the net effect is that gaseous absorption above the cloud seems to decrease both cloud absorptance and cloud reflectance by approximately the same amount.

The cloud models presented in Tables 2.1–2.3 show that liquid water content as well as cloud droplet distribution functions vary significantly from cloud top to cloud base. Observations by Mason (1971), Paltridge (1974b) and Stephens (1976) show that liquid water content is strongly variable with height within the cloud. Paltridge (1974b) also shows a sharp decrease in liquid water content near the cloud top, the region of maximum solar heating rates.

2.4.4 Vertically inhomogeneous clouds

Zdunkowski and Davis (1974) and Stephens (1976) modeled vertically homogeneous clouds in the infrared window region. These calculations show that the cloud radiation field and cooling rates are affected by the vertical distribution of the liquid water content. While Zdunkowski and Davis (1974) developed an analytic solution to variable liquid water contents in the solar spectrum, complete calculations have not previously been performed. Table 2.10 shows cloud heating rates, reflectance, transmittance and absorptance values (defined in terms of radiation incident upon the cloud top) for several vertical distributions of liquid water

Table 2.10. Heating rates, reflectance (R), transmittance (T) and absorptance (A) for a 300 m thick stratus cloud with cloud top at 1.6 km for various vertically inhomogeneous liquid water contents w_L (g m^{-3}).

	Case 1		Case 2		Case 3		Case 4		Case 5	
	Cloud top 1600 m w_L = 0.30 1500 m w_L = 0.20 1400 m w_L = 0.10 1300 m Cloud base		1600 m w_L = 0.30 1540 m w_L = 0.25 1480 m w_L = 0.20 1420 m w_L = 0.15 1360 m w_L = 0.10 1300 m		1600 m w_L = 0.20 1500 m w_L = 0.20 1400 m w_L = 0.20 1300 m		1600 m w_L = 0.10 1500 m w_L = 0.20 1400 m w_L = 0.30 1300 m		1600 m w_L = 0.30 1500 m w_L = 0.30 1400 m w_L = 0.30 1300 m	
Solar zenith angle	0°	60°	0°	60°	0°	60°	0°	60°	0°	60°
Average heating (°C h^{-1}) upper 100 m	1.25	0.47	1.18	0.45	0.86	0.36	0.54	0.25	1.14	0.44
Average total cloud heating (°C h^{-1})	0.59	0.20	0.60	0.20	0.61	0.20	0.61	0.21	0.74	0.23
R (%)	34.9	54.0	34.8	54.0	34.7	53.9	34.6	53.8	45.7	62.0
T (%)	60.9	43.0	61.0	42.9	60.9	43.0	61.0	43.0	49.1	34.5
A (%)	4.2	3.0	4.2	3.1	4.2	3.1	4.4	3.2	5.2	3.5

content (w_L). A cloud with thickness of 300 m is assumed with cloud top at 1600 m with solar zenith angles of $\theta = 0°$ and 60°.

Case 1 assumes three vertical layers, each 100 m thick with $w_L = 0.3$, 0.2 and 0.1 g m^{-3} varying from top to bottom. Case 2 assumes five vertical layers in order to determine if a three-layer model provides sufficient vertical resolution. Layers are 60 m thick, and liquid water contents vary from the top down as $w_L = 0.30$, 0.25, 0.20, 0.15 and 0.10 g m^{-3}, respectively. Within each layer average cloud attenuation parameters are assumed. Case 3 considers a uniform cloud with $w_L = 0.2$ g m^{-3} throughout. Case 4 considers liquid water contents of $w_L = 0.1$, 0.2 and 0.3 g m^{-3}, varying from top to bottom, and Case 5 considers a uniform cloud with $w_L = 0.3$ g m^{-3} throughout. In all cases the stratus top drop size distribution has been assumed.

Comparison of Cases 1 and 2 shows that the three-layer cloud gives results essentially identical to the five-layer cloud. Therefore, the division of the liquid water profile is not critical and 100 m thick layers appear sufficient for such radiation calculations.

Comparison of Cases 1, 3 and 4 shows that R, T and A and the average total cloud heating rate are almost identical for similar solar zenith angles. These results indicate that a uniform liquid water content throughout the cloud is adequate to determine cloud radiative characteristics. This conclusion holds for clouds of thicknesses up to at least 500 m. However, if detailed heating rates are required within the cloud boundary, then detailed layered calculations including liquid water variations with height are necessary. Comparison of heating rates averaged over the upper 100 m of the cloud shows that these values vary strongly with liquid water content. Case 5 shows once again that reflectance and absorptance increase with increasing liquid water content (optical depth) while transmittance decreases. However, note that the heating rate in the upper 100 m of the cloud is smaller in Case 5 than in Case 1. This is caused by a larger upward diffuse flux in the uniform w_L case (Case 5) than in the variable w_L case (i.e., less scattering). Since heating rates are a function of net flux divergence, an increase in upward diffuse flux will lead to smaller heating rates. Therefore, liquid water content can be important to heating rates even near the cloud top.

Table 2.11 shows values similar to those given in Table 2.10 for six different drop size distributions. Liquid water contents for Case 1 were assumed to decrease from $w_L = 0.3$ g m^{-3} at the cloud top to 0.1 g m^{-3} at cloud base; for Case 4 $w_L = 0.1$ g m^{-3} at cloud top, increasing to 0.3 g m^{-3} at cloud base. Cloud base heights of 1.3 and 9 km have been assumed with $\theta = 0°$. These results show that average total cloud heating rates and

TABLE 2.11. Heating rates, reflectance (R), transmittance (T) and absorptance (A) for Cases 1 and 4 as defined in Table 2.10 and for six different drop size distributions. Calculations are given for two different cloud base heights.

	Nimbostratus				Stratus				Stratocumulus			
	Top		Base		Top		Base		Top		Base	
	Case 1	Case 4	Case 1	Case 4	Case 1	Case 4	Case 1	Case 4	Case 1	Case 4	Case 1	Case 4
					a. Cloud base 1.3 km							
Average heating (°C h^{-1}) upper 100 m	1.04	0.46	1.39	0.95	1.25	0.86	1.52	0.64	1.10	0.48	1.56	0.63
Average total cloud heating	0.55	0.56	0.62	0.64	0.59	0.61	0.62	0.66	0.56	0.57	0.62	0.66
R (%)	25.8	25.7	40.2	40.1	34.9	34.6	47.3	47.2	28.5	28.3	45.5	45.3
T (%)	70.7	70.8	55.1	55.0	60.9	61.0	47.9	47.8	67.7	67.8	49.7	49.7
A (%)	3.5	3.5	4.7	4.9	4.2	4.4	4.8	5.0	3.8	3.9	4.8	5.0
	Case 1	Case 4	Case 1	Case 4	Case 1	Case 4	Case 1	Case 4	Case 1	Case 4	Case 1	Case 4
					b. Cloud base 9 km							
Average heating (°C h^{-1}) upper 100 m	6.1	2.9	7.9	3.8	7.3	3.4	8.9	4.2	6.4	3.1	8.6	4.2
Average total cloud heating	2.9	3.0	3.3	3.3	3.2	3.2	3.3	3.4	3.0	3.1	3.3	3.4
R (%)	25.1	24.9	39.1	39.0	34.9	34.8	46.3	46.2	28.6	28.5	44.5	44.3
T (%)	67.7	67.8	53.0	53.1	57.4	57.5	45.8	45.8	64.1	64.1	47.6	47.7
A (%)	7.2	7.3	7.9	7.9	7.7	7.7	7.9	8.0	7.3	7.4	7.9	8.0

cloud absorptance values are nearly identical for all assumed drop size distributions (as well as assumed vertical distributions of liquid water content). Therefore, it may be assumed that bulk cloud absorptance is primarily a function of total cloud liquid water content and not a function of how that liquid water is distributed vertically in the cloud. However, even assuming the same liquid water content, cloud reflectance and cloud transmittance are strongly dependent upon the drop size distribution. For these clouds, reflectance values may vary almost by a factor of 2 (26 to 47%) for the identical liquid water content. Comparing results for cloud base heights of 1.3 and 9 km shows again that cloud reflectance and transmittance values (defined as in Table 2.7) do not vary strongly with height. However, cloud absorptance nearly doubles in these two cases and average heating rates increase by a factor of 6.

Even though cloud-averaged heating rates do not vary significantly with either drop size distribution or variable liquid water content, that is not the case with local cloud heating rates within the cloud. The average heating rate in the upper 100 m of the cloud may vary by 25–30% with drop size distribution. However, it varies much more strongly with assumed vertical distribution of liquid water content. For Cases 1 and 4 in which liquid water content in the upper region varies from 0.3 to 0.1 g m^{-3} average heating rates in this region vary by up to a factor of 2.

Fig. 2.5 shows heating rates within the cloud for three liquid water content distributions defined in Table 2.10 as Cases 1, 2, 3. For uniform liquid water content (Case 3) heating rates decrease nearly linearly with depth with a maximum value of 1°C h^{-1} near cloud top. Case 1, which has a liquid water content maximum at cloud top, gives the largest values of heating, reaching 1.5°C h^{-1} in the top 100 m. However, heating

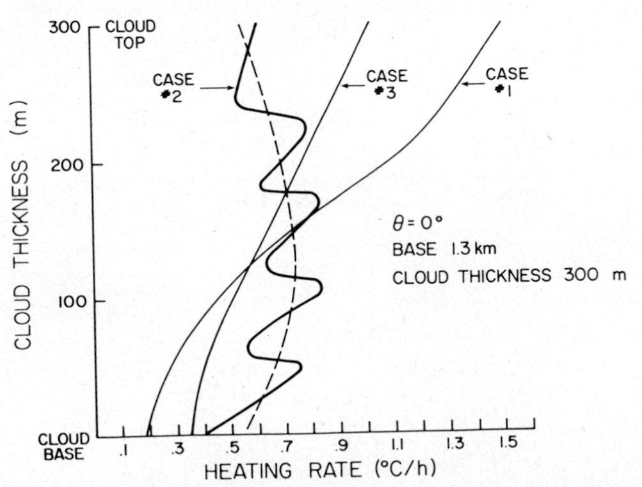

FIG. 2.5. Heating rates within a 300 m thick cloud for three liquid water content distributions, defined in Table 2.10.

rates in the lower portion of the cloud are smaller than for Case 3. Case 2 shows particularly interesting behavior. Within each layer, heating rates decrease as a function of depth within the cloud. However, at each layer interface, in which liquid water content is increased, there is a corresponding sudden increase in heating rates. Increasing the number of layers, so that a nearly uniform increase in liquid water content occurs with increasing depth, provides the heating rate curve shown by the dotted line in Fig. 2.5. In this case the cloud heating rate increases with optical depth in the upper half of the cloud with a maximum value found at a depth of 150–200 m below cloud top. It should be noted that these particular results apply only to a cloud 300 m thick. In any case, these results show that bulk and local radiative cloud properties may vary strongly with both liquid water contents and liquid water distributions. However, in light of results to be presented in Chapter 3 we must emphasize that these conclusions are based upon monomodal droplet size distributions.

2.4.5 Variations in cloud radiative characteristics as a function of drop size distribution for clouds of constant optical thickness

The previous sections have discussed the dependence of cloud bulk radiative characteristics upon droplet size distribution. The calculations upon which the analysis was based were performed as a function of cloud thickness for a constant drop concentration of $N = 100$ cm^{-3}. Calculations reported by McKee and Klehr (1977) indicate that cloud radiative properties at 0.5 μm do not vary strongly with cloud drop size distribution for clouds of equal optical thickness. Therefore, we shall examine other portions of the solar spectrum to see if they exhibit this same insensitivity to the droplet distributions.

Optical depth is a function of both extinction coefficient and cloud thickness. The extinction coefficient is a function of drop size distribution and drop concentration. Therefore, for a given drop size distribution, multiple combinations of drop concentration and cloud thickness can provide equal values of cloud optical depth. In the type of water clouds considered here, droplet concentrations may range from about 50 to 500 cm^{-3} (Mason, 1971; Pruppacher and Klett, 1978). While not necessarily an average value, concentrations of 100 cm^{-3} are often reported in the literature (Sartor and Cannon, 1977).

For the purposes of this section, cloud optical depth of approximately 50 has been chosen, with zenith angle of $\theta = 0°$; cloud top height has been kept at 2500 m in all cases. It should be recalled that for a given drop size distribution the value of extinction coefficient (β_e) varies with wavelength. As shown in Tables 2.2 and 3.2, the value of β_e may vary by as much as 20% across the solar spectrum, with values increasing with increasing wavelength. It is partially for this reason that we have for the most part avoided using the term optical depth in this monograph. We have, in this section, used a spectrally weighted average value of β_e to represent each drop size distribution.

Table 2.12a shows the values of cloud reflectance (R), transmittance (T) and absorptance (A) as a function of wavelength for several drop size distributions. The effect of varying geometric cloud thickness, keeping $N = 100$ cm^{-3} for each of the drop size distributions, is readily seen. The bulk radiative characteristics at most wavelengths as well as the total averaged characteristics are not significantly different for the different drop size distributions. However, the values of cloud reflectance near 1.8 μm do show rather strong variations with drop size distribution. Therefore, it is suggested that remote sensing in this region might provide information concerning the drop size distribution. The cloud heating rate responded in an expected manner being inversely proportional to the geometric thickness of the cloud layer.

Table 2.12b shows the results of varying droplet concentration while again keeping cloud optical depth constant. The first case consists of the stratus base distribution with drop concentration increased to $N = 400$ cm^{-3}, and a corresponding decrease in cloud thickness to 500 m. Comparing this case to the nimbostratus case having the same geometric thickness and optical depth shows that the stratus base distribution with a higher drop concentration has a larger value of reflectance, but significantly smaller values of cloud absorptance and average cloud heating rates. As a second comparison, the drop concentration for the nimbostratus top distribution is decreased to $N = 25$ cm^{-3} while cloud thickness is increased to 2000 m. This case, when compared with the stratus base case in Table 2.12a, has a smaller value of cloud reflectance but larger values of cloud absorptance and a correspondingly larger cloud average heating rate.

In each of the cases shown in Table 2.12a note that the value of cloud transmittance is a function of optical depth and not strongly dependent upon drop size distribution, drop concentration or cloud thickness.

Even though the two cases shown in Table 2.12b have equal values of cloud optical depth, their bulk radiative characteristics are significantly different. At wavelengths $\lesssim 0.8$ μm where there is very little absorption, bulk radiative characteristics are nearly independent of drop size distribution. As McKee and Klehr (1976) indicated, the value of cloud reflectance in this region appears to be only a function of optical depth. At wavelengths $\gtrsim 2.5$ μm, with almost total absorption, there is likewise almost no dependence upon drop size distribution. Drop size distribution is important only for absorption in the spectral region from 0.8 to 2.5 μm. Averaged over the entire solar spectrum, the following relationship holds: for a constant optical depth, the larger the droplet concentration, the larger the value of cloud reflectance, with correspondingly smaller values of cloud absorptance and cloud averaged heating rates. For optically thick clouds, the transmittance is virtually unaffected.

TABLE 2.12. Comparison of radiative properties of clouds with identical optical depths but different size distributions, geometric thicknesses and droplet concentrations. The R, T and A values are given in percentages.

TABLE 2.12a.

Drop size distribution	Cloud thickness (m)	Drop concentration (cm^{-3})		← Wavelength region (μm) →										Average cloud heating rate (°C h^{-1})
				0.55	0.765	0.95	1.15	1.4	1.8	2.8	3.35	6.3	Total	
Stratus base	2000	100	R	79.8	80.6	76.3	70.4	49.9	37.6	0.4	0.6	0.95	73.8	
			T	20.0	18.9	15.6	11.7	2.7	0.5	0.0	0.0	0.0	16.2	0.196
			A	0.2	0.5	8.1	17.9	47.4	61.9	99.6	99.4	99.05	10.0	
Nimbostratus top	500	100	R	79.1	79.3	77.5	74.1	40.3	23.1	0.2	0.3	0.4	72.5	
			T	20.6	19.9	18.0	15.3	1.7	0.2	0.0	0.0	0.0	17.3	0.863
			A	0.3	0.8	4.5	10.6	58.0	76.7	99.8	99.7	99.6	10.2	

TABLE 2.12b.

Drop size distribution	Cloud thickness (m)	Drop concentration (cm^{-3})		0.55	0.765	0.95	1.15	1.4	1.8	2.8	3.35	6.3	Total	Average cloud heating rate (°C h^{-1})
Stratus base	500	400	R	79.8	80.6	79.9	78.3	55.2	39.4	0.4	0.6	1.0	75.5	
			T	20.0	18.9	17.3	15.3	4.9	0.5	0.0	0.0	0.0	16.8	0.647
			A	0.2	0.5	2.8	6.4	40.9	60.1	99.6	99.4	99.0	7.7	
Nimbostratus top	2000	25	R	79.2	79.4	74.0	66.7	37.2	22.2	0.2	0.3	0.4	71.1	
			T	20.6	19.8	16.3	11.8	1.2	0.2	0.0	0.0	0.0	16.6	0.239
			A	0.2	0.8	9.7	21.5	61.6	77.6	99.8	99.7	99.6	12.3	

Note that the value of cloud reflectance is significantly different at 1.8 μm (Table 2.12a), while nearly the same at 0.95 and 1.15 μm, for drop size distributions with equal drop concentrations but different cloud thicknesses. On the other hand, there are large differences in the value of cloud reflectance even for the 0.95 and 1.15 μm wavelength regions for cases in which there are significantly different drop concentrations. Such differences suggest that remote sensing within the wavelength regions 0.80 to 2.5 μm may be able to discriminate between variations in drop concentrations and cloud thickness and, therefore, drop size distributions, even for clouds of equal optical thickness. Cloud optical thickness may be determined from the value of cloud reflectance at wavelengths <0.8 μm. This combination then prescribes a technique for remote sensing using the solar spectrum.

2.5 *Summary*

Since clouds play such a crucial role in the determination of the earth's energy budget, a rather thorough analysis of cloud reflectance (albedo), transmittance and absorptance under a wide variety of cloud conditions was presented. Since radiative-dynamical effects may be an important variable in cloud dynamics, an extensive study of cloud heating rates was undertaken. We showed that under identical cloud conditions absorption of solar radiation by clouds may increase markedly from clouds at low altitudes to clouds near the tropopause. A 1 km thick cloud layer absorbed nearly twice as much solar radiation when it was located at 10 km than the same layer would at an altitude of 1.5 km. Furthermore, height-dependent solar absorption was virtually unaffected by the drop size distribution. The water vapor path length above the cloud decreases rapidly with increasing cloud height; thereby allowing more interaction between the solar radiation and the cloud droplets and cloud water vapor. The combined effects of more energy in the water vapor absorption bands and a decrease in air density, explains the variations (by a factor of 4–6) in the heating rates which we calculated for low and high clouds.

For a liquid water content of 0.1 g m^{-3} in a 1 km thick cloud, average heating rates varied from about 0.3°C h^{-1} for a cloud top at 1.5 km to about 1.5°C h^{-1} for a cloud top at 10 km. Near cloud top extremely large heating rates occur, due primarily to droplet absorption. However, it was hypothesized that cloud inhomogeneities tend to distribute radiation absorption more uniformly throughout the cloud body than the present results suggest. The present model was based upon uniform cloud conditions throughout the cloud body. A study of radiation characteristics in a cloud with decreasing liquid water content from cloud top to cloud base showed that bulk cloud reflectance, transmittance and absorptance were nearly identical to those of a uniform cloud of equal optical depth. However, while cloud vertical structure may not be critical to the bulk cloud radiation field, it was important to local heating rates within the cloud.

Studies by McKee and Cox (1974, 1976) showed that horizontally finite clouds have smaller cloud directional reflectance values and larger cloud transmittance values. This is due to the fact that finite clouds "leak" radiation through their sides. A study by Davis *et al.* (1979a), however, showed that cloud absorptivity in finite clouds is less at small zenith angles and larger at large zenith angles when compared to absorptance in horizontally infinite clouds. Therefore, the cloud heating rates and absorptance values in the present investigation represent average conditions.

The present investigation also calculated integrated daily average heating rates in clouds in a tropical atmosphere. These integrated daily averages varied from about 0.1°C h^{-1} for a cloud top at 1.5 km to about 0.7°C h^{-1} for a cloud top at 10 km assuming a 1 km thick cloud and liquid water content of 0.1 g m^{-3}. We showed that integrated average daily heating rates in layer clouds may be approximated by a single calculation at a solar zenith angle of $\theta = 50°$ for a tropical atmosphere. However, this result is seasonally and latitudinally dependent.

The effect of various cloud droplet size distribution functions upon the cloud radiation field were also investigated. It was shown that cloud absorptance values are nearly independent of cloud droplet size distributions. However, both cloud reflectance (albedo) and transmittance values varied significantly with droplet size distribution. Absolute energy reflected from cloud tops was shown to be strongly dependent upon cloud top height and cloud optical depth as well as upon the drop size distribution. However, for a cloud of thickness of 300 m, vertical variations in liquid water content did not appear to significantly influence cloud reflectance.

In general, variations in cloud heating rates appeared to be mildly insensitive to seasonal or latitudinal variation in vertical water vapor profiles. Variations in heating rates between McClatchey's midlatitude summer, midlatitude winter and tropical water vapor profiles were less than 10%. However, cloud average heating rates using the oceanic GATE water vapor profile led to decreases of 20% compared to heating rates using the McClatchey tropical profile.

Measurements by Reynolds *et al.* (1975) for a 5 km thick cloud yielded an absorptance value of 36%. The present calculations gave an absorptance value of 17%, or approximately half that of observations, for the identical cloud conditions. These calculations were in agreement with Twomey (1976) and Stephens (1978) who demonstrated that maximum absorptance values in clouds were approximately 20%. In all previous theoretical cloud radiation studies, including the present chapter, monomodal drop size distributions have been assumed.

For clouds of spectrally averaged optical depths equal to 50, it was shown that the reflectance values for wavelengths less than 0.8 μm and greater than 2.5 μm were relatively independent of droplet size distribution and primarily dependent upon optical depth. However, near 1.8 μm, cloud reflectance was dependent upon droplet size distribution. These results suggest that one may be able to infer droplet size distribution functions from a set of observations in the 0.8 to 2.5 μm wavelength interval.

CHAPTER 3

The Effect of Cloud Bimodal Drop Size Distributions upon the Radiative Characteristics of Clouds

RONALD M. WELCH AND STEPHEN K. COX

Measurements by Reynolds *et al.* (1975) show that direct absorption of solar radiation in clouds is an important tropospheric heat source. Their observations show cloud absorptance values as high as 30–52%. For a particular case study the measurements of Reynolds *et al.* showed an absorptance value of 36% while calculations in Chapter 2 showed a corresponding absorptance value of 17%, i.e., less than half. Twomey (1976) has shown that theory predicts a maximum absorptance value of 20%. Moreover, Twomey (1972) showed that the addition of particulates does not add significantly to cloud absorption. Therefore, there has been a disparity between theory and observations. Twomey (1976) mentioned that in order to obtain such large absorptance values as measured by Reynolds *et al.* there must either be total absorption of solar radiation at wavelengths $> 0.7\ \mu m$ or significant absorption of solar radiation at wavelengths $< 0.7\ \mu m$. However, there has been no apparent mechanism postulated to satisfy either of these two conditions until now.

As mentioned above, all previous calculations have been concerned with monomodal drop size distributions. While these results have shown that the choice of drop size distribution is not important to the determination of cloud absorptance values, the inclusion of very large drops in a bimodal drop size distribution has not been considered. This is the topic of the present chapter.

3.1 Theoretical considerations

The method of calculation is identical to that presented in Chapter 2. The spherical harmonics method with the four-term expansion, developed by Zdunkowski and Korb (1974), has been converted into the delta-spherical harmonics technique. For large optical depths ($\tau > 35$), the adding method using the diamond initialization (Wiscombe, 1976) has been utilized. Intercomparisons of radiative fluxes and heating rates between these two techniques show excellent agreement. The vertical water vapor profile taken from GATE (Phase III) data has been used in most of the calculations. However, for low-lying clouds there may be differences as large as 20% in cloud absorptance values and heating rates between the GATE vertical water vapor profile and the tropical water vapor profile given by McClatchey *et al.* (1971).

3.1.1 THE PRESENCE OF LARGE DROP SIZES IN CLOUDS

There have been a large number of studies concerning microwave attenuation in clouds (Crane, 1977; Hogg and Chu, 1975; Medhurst, 1965; Semplak, 1970; Rogers, 1976). Even when there is no measured precipitation at the ground, these studies indicate the presence of large droplets in clouds from the strong attenuation of microwave radiation. Indeed, rain is prevalent in clouds up to an altitude of 10 km (Hogg and Chu, 1975). Therefore, it seems clear that a distribution of droplets with large radii is often present in thick developing clouds.

Mason and Jonas (1974) report that in maritime clouds containing small concentrations of droplets, the drop spectra broaden rapidly by condensation and produce droplets of $r = 25\ \mu m$ in concentrations of $10^{-3}\ cm^{-3}$ within half an hour. Larger droplets continue to grow rapidly by coalescence to precipitation size. Bimodality was found to be more pronounced at higher levels in the cloud and when the environment becomes increasingly unstable. These results were based on small cumuli of depth 2 km. Their model shows that successive cloud updrafts achieve increasing buoyancy, height, vertical velocity and liquid water content than their predecessors along with a broader droplet spectrum. Liquid water contents in this model double or even triple between successive updrafts with values greater than $2.5\ g\ m^{-3}$. It was also shown that a change in environmental humidity from 85 to 80% nearly halves the cloud liquid water content while it doubles when the humidity is raised to 90%. Liquid water contents were found to be strongly dependent upon lapse rate and the diameter of the updraft. Near cloud top, drops with $r \geq 20\ \mu m$ were predicted with concentrations of $10\ cm^{-3}$. This model shows that the bimodal droplet distributions develop in successive updrafts. Jonas and Mason

(1974) demonstrate that in small clouds the concentration of nuclei is a more important factor in controlling the onset of precipitation than is the updraft. They show that condensation can indirectly affect the growth rate of droplets $> 25\ \mu m$ by enhancing the growth of smaller drops which are then captured more efficiently by the large ones. The overall result is to produce a more rapid variation of the drop size distribution as well as a faster increase in the growth of large drops than would be the case by condensation and coalescence acting separately. However, the growth of droplets by coalescence depends strongly upon the collision efficiency (de Almeida, 1977) which may be altitude dependent.

Jonas and Mason (1974), Jonas and Goldsmith (1972) and Tennekes and Woods (1973) show that turbulence in clouds tends to accelerate the growth rate of large drops. However, Warner (1973, 1977) reports that changes in droplet spectrum and liquid water content vary little with cloud age in small cumuli. In addition, these observations indicate that the existence of an updraft somewhere within the cloud continues for at least a large fraction of cloud lifetime. Furthermore, liquid water content increases with height in a well-defined manner. The previously mentioned observations and models are associated with small cumuli. However, the large absorptance values observed by Reynolds *et al.* are typically associated with large cumuli. Whether or not the above models can be extended to large developing clouds is unknown. However, the models of Mason and Jonas (1974) and Jonas and Mason (1974) suggest that for massive updrafts of large vertical extent both liquid water content and drop concentrations may reach very large values.

Sartor and Cannon (1977) reported measurements of cloud microstructure in precipitating convective clouds. They found that the commonly held assumption that microphysical cloud properties are distributed randomly with respect to each other on the smaller scales may not be valid. Apparently the microstructure is not uniform throughout the cloud. Bimodal size distributions were measured in the middle and lower portions of the cloud. Most striking is their finding that observed frozen water content can increase by one to two orders of magnitude over liquid water content in a precipitation shaft. The observed change in ice particle concentration exceeded by two orders of magnitude the expected ice nuclei concentration usually found at comparable temperatures. The average concentrations of ice particles sometimes exceeded 400 ℓ^{-1}. Cloud droplet liquid water content in one study was measured to be 0.22 g m^{-3}; corresponding frozen water content in a precipitation shaft was measured to be ~ 7 g m^{-3}. Furthermore, they report that the presence of supercooled water droplets in a precipitation shaft may be depleted (98.5% complete) in as little as 60 s. It seems reasonable then to expect the presence of possibly large concentrations of large ice particles or water drops within convective clouds.

3.1.2 Calculation of bimodal attenuation parameters

In the presence of a bimodal drop size distribution function, attenuation parameters are additive, considering each drop size distribution separately. However, phase functions must be weighted by their scattering attenuation parameters (Deirmendjian, 1969) at each wavelength λ:

$$\overline{P(\cos\theta)} = \frac{P_1(\cos\theta)\,\beta_{s,1} + P_2(\cos\theta)\,\beta_{s,2}}{\beta_{s,1} + \beta_{s,2}}. \tag{3.1}$$

Here $P_1(\cos\theta)$ and $P_2(\cos\theta)$ are the individual phase functions for each drop size distribution, and $\beta_{s,1}$ and $\beta_{s,2}$ are the corresponding scattering parameters. $\overline{P(\cos\theta)}$ is the average, or bimodal phase function.

Drop size distributions with mode radii ranging from 6 to 600 μm have been used in the present investigation. For a drop of radius 1 mm (1000 μm) the size parameter at a wavelength of 0.35 μm is approximately 2×10^4. Exact Mie calculations for spherical water droplets have been carried out up to a size parameter $x = 6000$. For size parameters larger than 6000, approximate values have been used.

An approximate expression for the absorption efficiency is given by Van de Hulst (1957) and Deirmendjian (1969):

$$Q_a(\rho,m) = 1 + \frac{\exp(-2\rho\tan g)}{\rho\tan g} + \frac{\exp(-2\rho\tan g) - 1}{2(\rho\tan g)^2}, \tag{3.2}$$

where $\rho = 2x(n_r - 1)$ is Van de Hulst's normalized size parameter and $g = \tan^{-1}[n_i/(n_r - 1)]$ is the absorption parameter. Therefore, $\rho\tan g$ gives the energy absorbed along the path inside the particle.

A corresponding expression is used for the extinction efficiency:

$$Q_e(\rho,m) = 2 - \frac{4\cos g}{\rho}\exp(-\rho\tan g)\sin(\rho - g) + 4\left(\frac{\cos g}{\rho}\right)^2[\cos 2g - \exp(-\rho\tan g)\cos(\rho - 2g)]. \tag{3.3}$$

Although the expressions given in (3.2) and (3.3) show reasonable agreement with exact Mie calculations for $x \lesssim 6000$, the absorption efficiency was sometimes underestimated by up to 20%.

If the scattering coefficient for one of the size distributions is much larger (an order of magnitude) than the other, the phase function with the larger scattering coefficient will be dominant. In most bimodal drop size distributions there are significantly greater numbers of small droplets (by orders of magnitude) than of large drops. However, the large drops may contain (an order of magnitude) more liquid water content than the smaller droplets. Since scattering is primarily a function of droplet concentration rather than drop size, the scattering properties of bimodal cloud distributions are dominated by the smaller cloud droplets.

In Chapter 2 it was shown that droplet absorption parameters and the corresponding cloud absorptance values are nearly independent of droplet size distribution function. However, these results cannot be extrapolated to drops with large radii.

For very small size parameters ($x = 2\pi r/\lambda$), Rayleigh absorption efficiency is significant with respect to scattering efficiency (Deirmendjian, 1969). However, over a very broad range of size parameters, absorption efficiency of water and ice is much smaller (orders of magnitude) when compared to scattering efficiency. Much previous research has neglected droplet absorption using the argument that values of the imaginary component (n_i) of the complex index of refraction ($m = n_r - in_i$) for water and ice across most of the solar spectrum are small. However, Welch *et al.* (1976) have shown that droplet absorption (for wavelengths $\gtrsim 2.5$ μm) can be significant when considering absorptance values and heating rates. This is due to the fact that within these spectral regions, large droplet absorption coefficients ($\beta_a \approx \beta_s$) lead to almost complete absorption of solar radiation.

Deirmendjian (1969) points out that absorption efficiency approaches scattering efficiency for very large size parameters, *no matter how weak the absorption properties* of the material in interaction with the radiation field, i.e.,

$$\lim_{x\to\infty} Q_a \approx \lim_{x\to\infty} Q_s = 1,\ n_i \ll 1.$$

Therefore, distributions of large drops may provide large absorptance values over a larger portion of the solar spectrum.

Chylek (1977) has shown that

$$\lim_{x\to\infty} Q_s(m,x) = 1 + \left|\frac{m-1}{m+2}\right|^2. \tag{3.4}$$

However, for ice water, $m \approx 1.33$; thus the conclusion that large particle sizes may provide large values of absorption remains unaltered.

Pruppacher and Pitter (1971) have studied the shape of water drops falling at terminal velocity with radii between 170 and 4000 μm. Their studies show that drops with radii < 170 μm are slightly deformed but may be considered spherical; the shape of drops between about 170 and 500 μm can be closely approximated by an oblate spheroid; drops between 500 and 2000 μm are deformed into asymmetric oblate spheroids with increasingly pronounced flat bases; and drops $\gtrsim 2000$ μm develop a concave depression in the base. Jones (1959) showed that the instantaneous shapes of natural raindrops varied from slightly prolate to oblate ellipsoids for a given drop size and that natural raindrops seem to be typically in a state of oscillation. However, due to the lack of a sufficient number of samples no average conditions were determined. Oguchi (1975) has considered scattering of microwave radiation by raindrops with a flat base and hemispherical top. These studies indicate that the shape factor is significant for the prediction of polarization effects but not for the determination of attenuation when compared to spherical drops. Furthermore, such effects decrease rapidly with increasing frequency. In addition, the presence of a broad drop size distribution tends to average out many of the nonspherical effects. Rogers (1976) concludes that the effects of drop deformation upon attenuation parameters are less important than those due to the normal variability of the drop size distribution. Therefore, nonspherical corrections to Mie theory are neglected in the results presented in this chapter.

The complex indices of refraction for water at wavelengths > 0.7 μm are based upon the measurements of Irvine and Pollack (1968). Values below 0.7 μm are taken from *Linkes Meteorologisches Taschenbuch* II (1953). For each spectral interval used in the present investigation, Zdunkowski *et al.* (1967) have provided the complex index of refraction, $m = n_r - in_i$, weighted by the solar blackbody distribution function at 6000 K, i.e.,

$$\overline{m} = \frac{\int_{\lambda_1}^{\lambda_2} m_\lambda B_\lambda\,(T = 6000\ \text{K})d\lambda}{\int_{\lambda_1}^{\lambda_2} B_\lambda\,(T = 6000\ \text{K})d\lambda}. \tag{3.5}$$

3.1.3 Large-particle mode size distributions

A number of drop size distributions with differing mode radius r_c have been used in the present investigation. All of these distributions are based upon the modified gamma distribution function (Deirmendjian, 1969)

$$n(r) = a\, r^\alpha \exp\left[-\frac{\alpha}{\gamma}\left(\frac{r}{r_c}\right)^\gamma\right], \tag{2.3}$$

expressed in cm^{-3} μm^{-1}.

Table 3.1 gives the values of a, α, γ, r_c, liquid water content (w_L) and droplet concentration (N) for each of the drop size distributions considered in the present

TABLE 3.1. Modified gamma drop size distribution parameters, liquid water contents (w_L) and drop concentrations (N) for the cases considered in the present investigation. Values are taken from Deirmendjian (1969, 1975). (Notation, e.g., 0.500 − 03 = 0.500*10⁻³).

Distribution	a	α	γ	r_c (μm)	w_L (g m^{-3})	N (cm^{-3})
C.5	0.5481	4	1.0	6.0	0.297	10^2
C.6	0.500 − 03	2	1.0	20.0	0.251	10^0
Rain L	0.497 − 07	2	0.5	70.0	0.117	10^{-3}
Rain L′	0.497 − 06	2	0.5	70.0	1.170	10^{-2}
Rain 10	1.033 − 14	4	1.0	333.3	0.509	10^{-3}
Rain 50	1.388 − 20	6	1.0	600.0	2.110	10^{-3}

investigation. The C.5 drop size distribution function (Deirmendjian, 1975) is representative of nimbostratus clouds and is based upon the measurements of Okita (1961). This distribution coincides almost exactly with measured distributions for cumulus congestus and nimbostratus clouds (Hansen, 1971). Therefore, the C.5 distribution was adopted as the "typical" small droplet size distribution function in the present investigation.

The C.6, Rain L, Rain 10 and Rain 50 distributions are representative of the large drop sizes found in bimodal size distributions. Distribution C.6 is representative of large droplet spectra (r_c = 20 μm) often found in precipitating clouds. The Rain L distribution (r_c = 70 μm) is representative of light rain. The Rain 10 distribution (r_c = 333 μm) represents moderate rainfall with a precipitation rate of 10 mm h^{-1} measured at the ground. The Rain 50 distribution (r_c = 600 μm) represents heavy rain with rainfall rates of 50 mm h^{-1} measured at the ground.

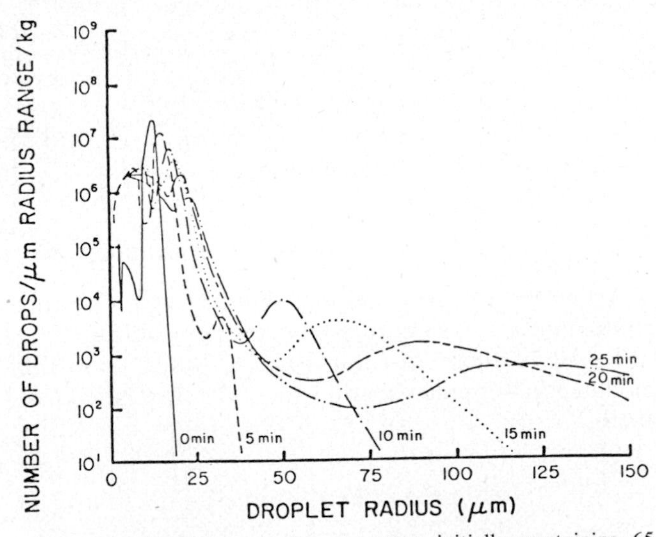

FIG. 3.1. Development of a spectrum initially containing 65 droplets mg^{-1} by condensation and coalescence in the maritime cumulus of Mason and Jonas. The collection kernels given in Fig. 1 of Jonas and Mason 1974 (p. 289) were assumed together with an entrainment parameter of 1.5 × 10^{-3} s^{-1} and a supersaturation of 0.25%. Spectra are shown at 5 min intervals. (From Jonas and Mason 1974, p. 291.)

These distribution functions are only representative cases and should be used with caution. The choice of particle distribution is arbitrary, no matter which is used. The small cumulus cloud model developed by Jonas and Mason (1974) indicates that drop spectra may vary substantially in as little as 5 min (Fig. 3.1). In fact, a trimodal drop spectrum is shown by Jonas and Mason with small size drops ($r \approx$ 5–10 μm), intermediate sized drops ($r \approx$ 15–25 μm) and large sized drops (with a variety of large drop sizes). One striking aspect of this figure is the nearly flat drop size spectrum calculated for the large drops. Concentrations of 10 cm^{-3} were reported for drops with radii $r \gtrsim$ 20 μm, or one order of magnitude larger than assumed for the C.6 distribution. Furthermore, concentrations of 10^{-4} cm^{-3} are reported for drops of r = 100 μm after only 20–30 min. Due to the broad drop spectrum, the possibility exists that much larger drop concentrations and liquid water contents may exist than represented by the idealized drop distribution functions assumed in the present investigation. Therefore, drop concentrations will be scaled upward in a later section of this chapter to determine if this might significantly affect the cloud's radiative characteristics. The previously mentioned results of Mason and Jonas (1974), Jonas and Mason (1974) and Warner (1973, 1977) are for small cumulus clouds. For the large cumulus clouds considered in the present investigation, estimates of liquid water content and drop concentrations may be significantly underestimated. It should be pointed out that Warner's (1973) results show that the drop size distribution changes little with cloud age. However, Warner's results are primarily concerned with drops of diameters in the 7–20 μm range. His results do not consider precipitation-sized particles or larger cloud droplets which are growing by coalescence. No particle concentrations less than 0.01 cm^{-3} were measured. However, as will be seen in later sections large drop concentrations of 10^{-3} cm^{-3} significantly affect cloud radiative characteristics, particularly absorption.

3.1.4 LARGE PARTICLE EXTINCTION AND ABSORPTION COEFFICIENTS

Table 3.2 gives extinction (β_e) and absorption (β_a) coefficients (km^{-1}) along with the drop absorption ratio ($k = \beta_a/\beta_e$) and the C_1 phase function expansion coefficients for the various drop size distributions.

Comparison of the C.5 distribution function and other distributions with mode radii ranging from 5 to 10 μm are given in Tables 2.1–2.3 of Chapter 2. Cloud absorptance values and heating rates were shown to be insensitive to the actual distribution function used assuming a monomodal drop size dis-

TABLE 3.2. Extinction (β_e) and absorption (β_a) coefficients (km^{-1}) along with drop absorption ratios ($k = \beta_a/\beta_e$) and the C_1 phase function expansion coefficients for the drop size distributions used in the present investigation.

Drop size distribution		Wavelength region (μm)									
		0.35	0.55	0.765	0.95	1.15	1.40	1.85	2.80	3.35	6.3
C.5	β_e	42.82	44.28	44.65	45.17	45.41	45.95	46.55	47.97	48.61	54.51
	β_a	9.56–05	1.64–05	1.08–03	8.27–03	2.84–02	0.41	1.12	23.61	23.49	20.69
	k	2.2–06	3.7–07	2.4–05	1.8–04	6.3–04	8.8–03	0.02	0.49	0.48	0.38
	C_1	2.59	2.59	2.59	2.57	2.55	2.55	2.53	2.88	2.80	2.80
C.6	β_e	7.64	7.66	7.69	7.74	7.75	7.79	7.83	7.89	7.94	8.15
	β_a	7.88–05	1.28–05	8.52–04	6.36–03	1.73–02	0.29	0.76	3.78	3.69	4.00
	k	1.03–05	1.67–06	1.1–04	8.2–04	0.002	0.037	0.097	0.48	0.46	0.49
	C_1	2.63	2.64	2.64	2.63	2.63	2.64	2.68	2.93	2.87	2.93
Rain L	β_e	0.366	0.366	0.366	0.366	0.366	0.366	0.367	0.367	0.368	0.371
	β_a	2.93–05	5.24–06	3.46–04	2.59–03	7.06–03	0.084	0.136	0.173	0.168	0.176
	k	8.00–05	1.43–05	9.46–04	7.1–03	0.019	0.23	0.37	0.47	0.46	0.47
	C_1	2.57	2.54	2.53	2.57	2.64	2.69	2.79	2.94	2.89	2.94
Rain 10	β_e	1.30	1.30	1.30	1.30	1.30	1.30	1.30	1.31	1.31	1.32
	β_a	1.22–04	2.27–05	1.49–04	0.011	0.031	0.357	0.543	0.614	0.597	0.624
	k	9.39–05	1.75–05	1.1–04	8.5–03	0.024	0.27	0.42	0.47	0.46	0.47
	C_1	2.56	2.55	2.52	2.50	2.56	2.69	2.77	2.91	2.84	2.95
Rain 50	β_e	3.52	3.52	3.52	3.53	3.53	3.53	3.54	3.54	3.55	3.56
	β_a	5.02–04	8.60–05	5.59–03	0.044	0.120	1.21	1.59	1.67	1.62	1.69
	k	1.43–04	2.44–05	1.6–03	0.012	0.034	0.34	0.45	0.47	0.46	0.47
	C_1	2.56	2.61	2.52	2.54	2.62	2.65	2.74	2.90	2.79	2.94
Rain L′	β_e	3.66	3.66	3.66	3.66	3.66	3.66	3.67	3.67	3.68	3.71
	β_a	2.93–04	5.24–05	3.46–03	2.59–02	7.06–02	0.84	1.36	1.73	1.68	1.76
	k	8.00–05	1.43–05	9.45–04	7.1–03	0.019	0.23	0.37	0.47	0.46	0.47
	C_1	2.57	2.54	2.53	2.57	2.64	2.69	2.79	2.94	2.89	2.94

tribution. However, as reported in Chapter 2, cloud reflectance and transmittance values do depend strongly upon the choice of droplet size distribution function.

Table 3.2 shows the extinction coefficients (β_e) are relatively independent of wavelength. However, absorption coefficients vary strongly as a function of both wavelength and drop size distribution. The spectral bandwidths are identical to those given in Welch *et al.* (1976). Variations in absorption properties for the various drop size distributions may be most clearly seen from the drop absorption ratio k. The C.5 distribution shows that droplet absorption efficiency is approximately equal to droplet scattering efficiency for wavelengths $\gtrsim 2.5$ μm. In these regions one expects almost total absorption of solar radiation. However, absorption in the 1.85 and 1.40 μm spectral band regions is also significant.

The drop absorption ratio increases rapidly at all wavelengths beyond 2.5 μm for an increase in drop mode radius from 6 to 20 μm for both the C.5 and C.6 distributions. A further increase to a drop mode radius of 70 μm, Rain L, shows significant drop absorption ($k \gtrsim 10^{-3}$) down to wavelengths of 0.7 μm. However, for further increases in drop size mode radius, i.e., 333 μm for Rain 10 and 600 μm for Rain 50, the increase in drop absorption ratio is relatively small. Therefore, it would appear that a bimodal drop size distribution with the large drops represented by Rain L, or some other comparable distribution, may be sufficient to produce the very large absorptance values observed in clouds (Reynolds *et al.*, 1975). Variations in drop absorption ratios (k) are relatively small between the Rain L, Rain 10 and Rain 50 distribution, even though these distributions are quite dissimilar and vary by almost an order of magnitude in drop mode radius. This would tend to indicate that for very large drops, attenuation parameters are primarily a function of liquid water content, or drop concentration, rather than the form of the distribution function.

The Rain L′ distribution consists of the Rain L distribution with the drop concentrations scaled arbitrarily upwards by a factor of 10. The variation in the attenuation parameters shown in Table 3.2 for the Rain L, Rain 10 and Rain 50 distributions is largely a result of variations of liquid water contents (0.12, 0.51 and 2.11 g m^{-3}, respectively). However, these results show that scaling of attenuation coefficients by liquid water content is not generally valid. Linearly increasing the attenuation coefficients of Rain L by a factor of 18, the ratio of liquid water contents of the Rain 50 and Rain L distributions, gives extinction values nearly double those of the Rain 50 distribution. In the spectral regions with wavelengths > 1.40 μm, the absorption coefficients also cannot be scaled according to liquid water content. However, the absorption coefficients do seem to be nearly linear with increasing liquid water content in the regions of small drop absorption ratio (i.e., $k \lesssim 0.01$). Scattering coefficients may be found by taking the difference between the extinction and absorption coefficients.

From Eq. (3.1) and Table 3.2, it is obvious that the scattering phase function in a bimodal drop size distribution is determined almost exclusively from the small particle size distribution. Even though the Rain 50 distribution contains approximately seven times as much liquid water as the C.5 distribution, the scattering coefficients for C.5 are more than an order of magnitude larger than those for Rain 50.

Jonas and Mason (1974) suggest in their model that the concentration of particles with $r \gtrsim 20$ μm may be 10 cm^{-3}. The attenuation coefficients calculated using the C.6 distribution must be scaled upward by an order of magnitude in order to represent this case. The consequences of scaling the drop concentrations upward to meet these conditions will be discussed later.

3.1.5 Phase functions

Table 3.2 shows the C_1 phase function expansion coefficient for each of the size distribution functions. Generally, phase functions are not supplied to the reader, due to the large space such tables require. However, only the C_1 expansion coefficient is required in calculations using the delta-eddington method (Joseph *et al.*, 1976). The success of this technique, as well as the four-term delta-spherical harmonics method used in the present investigation, stems from the fact that radiative fluxes are most strongly dependent (Wiscombe, 1977; and Chapter 2) upon the first few terms of the phase function expansion

$$P(\cos\theta) = \sum_{\ell=0}^{N} C_\ell P_\ell(\cos\theta), \tag{3.7}$$

where C_ℓ are the phase function expansion coefficients and $P_\ell(\cos\theta)$ the Legendre polynomials.

The asymmetry factor g is defined as

$$g = \langle\cos\theta\rangle = \int_\Omega P(\cos\theta)\, d\Omega = C_1/3, \tag{3.8}$$

where $d\Omega$ is an element of solid angle. It is well known that the Henyey-Greenstein phase function,

$$P(\cos\theta) = \sum_{\ell=0}^{N} (2\ell+1) g^\ell P_\ell(\cos\theta), \tag{3.9}$$

produces relatively accurate determinations of radiative fluxes and only requires knowledge of the C_1 expansion coefficient. Wiscombe (1977) shows that applying the delta function approximation leads to the form

$$P(\cos\theta) = 2f\,\delta(1-\cos\theta) + (1-f)\sum_{\ell=0}^{\infty} C'_\ell P_\ell(\cos\theta). \tag{3.10}$$

The four-term spherical harmonics technique developed by Zdunkowski and Korb (1974) can be easily converted into the delta-spherical harmonics technique with the replacement of

$$\tau \to \tau',$$

$$k \to k',$$

$$C_\ell \to C'_\ell,$$

where

$$\tau' = (1-\tilde{\omega} f)\,\tau,$$

$$k' = k/(1-\tilde{\omega} f),$$

$$C'_\ell = (2\ell+1)\chi'_\ell,$$

and

$$\chi'_\ell = \frac{C_\ell/(2\ell+1) - f}{1-f}$$

or

$$\chi'_\ell = \frac{g^\ell - f}{1-f} \text{ (in the Henyey-Greenstein form)},$$

and $\tilde{\omega}$ is the single-scattering albedo; f represents the fraction of the radiative energy remaining in the diffraction peak which does not scatter during an interaction between an electromagnetic wave and a particle. Wiscombe (1977) shows that

$$f = g^2$$

for the delta-eddington model. However, for the four-term delta-spherical harmonics technique, $f = C_4/9$ (or $f = g^4$ in the Henyey-Greenstein form). With only the first term C_1 in the phase function expansion series, it is possible to construct accurate radiative flux calculations.

3.2 *Radiative transfer calculations*

3.2.1 Monomodal large drop size distributions in thin clouds

Table 3.3 gives cloud reflectance *(R)*, transmittance *(T)* and absorptance *(A)* values for various spectral band regions. These values are given for the C.5, C.6, Rain L, Rain L′, Rain 10 and Rain 50 drop size distributions for a cloud 500 m thick with cloud top at 1.0 km. The GATE water vapor profile (see Chapter 2) was used. The solar zenith angle is assumed to be $\theta = 0°$ with a surface albedo $A_s = 0.04$ (sea surface). The results in Table 3.3 assume monomodal drop size distribution functions. Therefore, the results presented assume that the cloud is entirely composed of large drops for the Rain L, Rain L′, Rain 10 and Rain 50 size distributions. While these cases are actually representative of rain showers, it is useful to review the relative contributions to the radiation field from each distribution separately before combining them to form

TABLE 3.3. Percent cloud reflectance (*R*), transmittance (*T*) and absorptance (*A*) values for the C.5, C.6, Rain L, Rain L′, Rain 10 and Rain 50 drop size distributions as a function of spectral region. Cloud thickness is 500 m with cloud top at 1 km. Solar zenith angle θ = 0°.

Drop size distribution		← Wavelength region (μm) →									
		0.55	0.76	0.95	1.15	1.4	1.8	2.8	3.3	6.3	Total
C.5	*R*	62.3	62.7	61.5	57.5	41.0	30.0	0.2	0.4	0.6	59.0
	T	37.7	37.2	33.3	28.6	14.9	7.2	0.0	0.0	0.0	33.5
	A	0.0	0.1	5.2	13.9	44.1	62.8	99.8	99.6	99.4	7.5
C.6	*R*	4.9	4.9	4.6	4.2	3.5	3.7	1.1	1.1	0.6	4.6
	T	95.1	95.1	92.7	88.4	83.4	86.9	58.1	56.7	46.5	93.0
	A	0.0	0.005	2.7	7.4	13.1	9.4	40.8	42.2	52.9	2.4
Rain L	*R*	4.2	4.2	4.0	3.5	3.0	3.3	1.6	1.5	0.9	4.0
	T	95.8	95.8	93.3	89.3	84.7	89.0	67.5	65.8	54.5	93.8
	A	0.002	0.003	2.7	7.2	12.3	7.7	30.9	32.7	44.6	2.2
Rain L′	*R*	6.6	6.6	6.0	5.0	3.2	2.4	0.8	0.8	0.5	6.0
	T	93.4	93.4	90.9	86.9	76.2	72.8	48.9	47.9	39.1	90.4
	A	0.02	0.04	3.1	8.1	20.6	24.8	50.3	51.3	60.4	3.6
Rain 10	*R*	6.2	6.4	6.1	5.0	2.6	2.1	0.8	0.8	0.5	5.7
	T	93.8	93.5	90.5	86.1	71.8	69.3	50.6	49.3	41.2	90.2
	A	0.001	0.08	3.4	8.9	25.6	28.6	48.6	49.9	58.3	4.1
Rain 50	*R*	9.7	11.4	9.9	7.1	1.9	1.0	0.3	0.5	0.2	8.9
	T	90.3	88.2	84.5	79.0	45.5	39.3	28.5	27.1	23.6	83.4
	A	0.006	0.4	5.6	13.9	52.6	59.7	71.2	72.4	76.2	7.7

bimodal distributions. Wavelength band regions are those given by Welch *et al.* (1976); and 0.76 μm region extends from 0.70 to 0.80 μm; and the 0.55 μm regions is an average for all wavelengths < 0.70 μm.

Table 3.3 shows that the reflectance of the large drop distributions is much smaller than that of the small drop distribution (C.5) for wavelengths $\lesssim 2.5$ μm. This illustrates that scattering is more sensitive to the larger numbers of small particles than it is to particle size. However, for wavelengths $\gtrsim 2.5$ μm, reflectance is small in all cases due to the large-droplet absorption.

For the total solar spectral region Table 3.3 shows that the large-particle distributions have a reflectance value an order of magnitude smaller than the small-particle distribution. Although not shown in Table 3.3, the cloud absorptance varies from 2.5 to 8% for liquid water contents varying by an order of magnitude. Increasing the particle concentration of Rain L by an order of magnitude (Rain L′) has surprising little impact upon cloud radiative characteristics.

However, the results given in Table 3.3 are for a thin cloud with low base height. In Chapter 2 it was shown that increasing cloud height increases cloud absorptance. Table 3.4 shows corresponding values of *R*, *T* and *A* (for the entire solar spectrum) as a function of cloud thickness and cloud height.

Comparing the C.5 drop size distribution in Table 3.3 with the first column in Table 3.4 shows that increasing cloud thickness from 500 m to 1 km increases cloud reflectance from 59 to 71%; cloud absorptance increases from 7.5 to 10%. Further increases in cloud thickness to 2 and 3 km show that cloud reflectance is rapidly approaching an asymptotic value. However, cloud absorptance continues to increase with increasing cloud depth.

As discussed in Chapter 2, increasing base height from 500 m to 5 km did not change cloud reflectance significantly. However, cloud absorptance increased strongly when the cloud base height was raised. Simi-

TABLE 3.4. Percent cloud reflectance (*R*), transmittance (*T*) and absorptance (*A*) for the C.5, C.6, Rain L, Rain L′, Rain 10 and Rain 50 drop size distributions with values given for the entire solar spectrum. Solar zenith angle θ = 0°.

Drop size distribution		Cloud base (m) 500	500	500	5000	5000	5000
		Cloud thickness (m) 1000	2000	3000	500	2000	3000
C.5	*R*	71.2	77.8	79.8	57.1	74.5	76.2
	T	18.7	9.5	6.1	31.8	8.7	5.6
	A	10.1	12.7	14.1	11.1	16.8	18.2
C.6	*R*	5.5	7.3	9.4	4.6	7.1	9.1
	T	89.9	84.1	79.1	92.2	82.5	77.2
	A	4.6	8.6	11.5	3.2	10.4	13.7
Rain L	*R*	4.1	4.3	4.6	3.9	4.2	4.4
	T	91.7	88.0	85.2	93.3	86.8	83.8
	A	4.2	7.7	10.2	2.8	9.0	11.8
Rain L′	*R*	8.5	13.7	18.8	5.7	12.8	17.5
	T	84.7	74.1	65.1	89.2	71.6	62.4
	A	6.8	12.2	16.1	5.1	15.6	20.1
Rain 10	*R*	7.8	12.4	16.9	5.4	11.6	15.8
	T	85.6	74.1	65.5	88.8	71.0	62.1
	A	7.6	13.5	17.6	5.8	17.4	22.1
Rain 50	*R*	14.7	24.9	32.0	8.3	22.9	29.4
	T	72.4	55.3	44.0	80.3	51.3	40.3
	A	12.9	19.8	24.0	11.4	25.8	30.3

lar behavior is observed for the large drop size distributions. The C.6 drop size distribution shows radiative cloud characteristics similar to those of the large drop distributions due to the fact that the C.6 particle concentration is relatively small. The small values of cloud reflectance once again illustrate the fact that cloud scattering properties are more dependent on droplet concentrations than on particle size. It is interesting to note that the C.5 small-particle size distribution provides much smaller absorption coefficients than the large drop modes over much of the solar spectrum; nevertheless, in thin clouds the small-particle mode absorbs more radiation than most of the large particle modes. This behavior may be interpreted as multiple scattering's contribution to particle absorption; the more light-particle interactions, the larger the particle absorption. As cloud optical depth increases the large-particle modes develop much larger scattering fields (and cloud reflectance) along with large absorptance values.

Note that for a cloud thickness of 2 km, invariant of cloud base height, cloud absorptance values are similar for both the small-particle (C.5) and large-particle (Rain L′ and Rain 10) modes. However, the Rain 50 mode already has a cloud absorptance value of 20%, the asymptotic value of the C.5 distribution.

For a cloud thickness of 3 km the large-particle modes provide significantly larger values of cloud absorptance than do the small-particle modes. As in the case for the small-particle modes, increasing base height from 500 m to 5 km increases cloud absorptance values dramatically. The Rain 50 size distribution shows cloud absorptance values ranging between 20 and 30%, for cloud thickness of 2–3 km. Such absorptance values are within the range of values reported by Reynolds *et al.* (1975). Therefore, it appears that monomodal distributions of large particles may be able to provide the large values of absorptance shown by observations.

3.2.2 Monomodal large drop size distributions in thick clouds

The results presented in Tables 3.3 and 3.4 were for clouds of thin and moderate thickness. However, the results of Reynolds *et al.* were for very deep clouds, often extending from near the surface to near the tropopause. In order to explore the radiative characteristics of such thick clouds, Table 3.5 shows calculations for a 9 km thick cloud with base height at 500 m. Solar zenith angle is $\theta = 0°$.

For the C.5 small-particle size distribution function, cloud absorptance is about 21%, in agreement with the results in Chapter 2 and with Twomey (1976). The C.6 distribution produces a value of cloud absorptance of about 24% while the Rain L gives only a value of 20%. Increasing the Rain L concentration by an order of magnitude (Rain L′), however, increases cloud absorptance to 30%, comparable to that obtained using the Rain 10 distribution. Most striking is the fact that the Rain 50 size distribution produces a cloud absorptance value of 39%, or nearly double of the maximum found for small-particle distributions (Chapter 2). Such large values are also comparable to values

Table 3.5. Percent cloud reflectance (*R*), transmittance (*T*) and absorptance (*A*) for the C.5, C.6, Rain L, Rain L′, Rain 10 and Rain 50 drop size distributions as a function of spectral region. Cloud thickness is 9 km with cloud top at 9.5 km. Solar zenith angle $\theta = 0°$.

Drop size distribution		Wavelength region (μm) 0.55	0.76	0.95	1.15	1.4	1.8	2.8	3.3	6.3	Total
C.5	*R*	96.9	95.5	82.7	75.9	40.1	26.1	0.2	0.4	0.6	77.5
	T	3.1	2.4	0.3	0.0	0.0	0.0	0.0	0.0	0.0	1.6
	A	0.04	2.1	17.0	24.1	59.9	73.9	99.8	99.6	99.4	20.9
C.6	*R*	27.3	27.3	21.9	18.2	7.0	4.0	0.1	0.2	0.1	20.6
	T	72.7	72.6	58.2	49.2	22.7	16.6	0.8	1.4	0.6	55.7
	A	0.01	0.1	19.9	32.6	70.3	79.4	99.1	98.4	99.3	23.7
Rain L	*R*	9.1	9.2	6.7	4.8	1.7	1.0	0.2	0.3	0.1	6.5
	T	90.9	90.7	75.6	67.3	36.7	33.5	13.0	22.9	11.3	73.1
	A	0.04	0.1	17.7	27.9	61.6	65.5	86.8	76.8	88.6	20.4
Rain L′	*R*	51.5	51.7	37.8	25.4	3.3	0.8	0.1	0.2	0.1	35.7
	T	47.8	47.2	36.2	30.3	2.6	0.5	0.0	0.0	0.0	33.9
	A	0.07	1.1	26.0	44.3	94.1	98.7	99.9	99.8	99.9	30.4
Rain 10	*R*	47.6	48.4	35.1	22.1	1.5	0.5	0.1	0.3	0.1	32.8
	T	52.3	49.2	32.7	23.6	1.0	0.2	0.1	0.1	0.1	34.6
	A	0.05	2.4	32.2	54.3	97.5	99.3	99.8	99.6	99.8	32.6
Rain 50	*R*	69.5	68.2	38.9	20.6	1.5	0.5	0.1	0.4	0.1	44.5
	T	31.4	22.4	7.2	2.4	0.0	0.0	0.0	0.0	0.0	16.5
	A	0.1	10.4	53.9	77.0	98.5	99.5	99.9	99.6	99.6	39.0

measured by Reynolds *et al.* However, it should be mentioned that in one case a cloud absorptance value of 52% was measured. The present results indicate that even in precipitation shafts of extremely thick clouds one would not expect absorptance values $\gtrsim$ 40%. The C.6 distribution has a small liquid water content and particle concentration. Increasing the C.6 concentration by an order of magnitude increases cloud reflectance to 60% and absorptance to 28.7%, values intermediate between those obtained from the small-particle (C.5) and large-particle distributions. Increasing the concentration of small droplets (in these very thick clouds) leads to a slight increase in cloud reflectance and a slight decrease in cloud absorptance. At this point, however, it is not clear what the effect of a bimodal drop size distribution has upon the cloud radiative characteristics. Of particular note, however, is the fact that frozen water contents of 7 g m^{-3} were reported by Sartor and Cannon for convectively active clouds of moderate thickness. It will be shown in the following chapter that ice particles have radiative characteristics surprisingly similar to water droplets in the solar spectrum. Therefore, the presence of large amounts of either ice or water has significant implications upon measured or calculated values of cloud absorptance.

The Rain L′ size distribution, while having a liquid water content of nearly twice that of the Rain 10 distribution, provides a smaller cloud absorptance than does the Rain 10 distribution. Therefore, drop size clearly is a more important factor for cloud absorptance than is liquid water content.

Table 3.5 shows that at wavelengths $\gtrsim$ 1.3 μm, almost all the solar radiation entering the cloud is absorbed for the Rain 50, Rain 10 or Rain L′ size distributions. However, more than 50% of the solar radiation is absorbed for the large-drop Rain 50 distribution at wavelengths > 0.8 μm, and 10% of the solar radiation absorbed down to 0.70 μm. The impact of the large droplets seems to be concentrated on wavelengths between 0.70 μm and 1.3 μm. Comparison of values in this spectral region in Table 3.5 between the Rain L′, Rain 10 and Rain 50 distributions demonstrates this effect most clearly. However, even for the very large drop sizes there is effectively no significant absorption at wavelengths < 0.70 μm.

In response to Sartor and Cannon's report of such large values of frozen liquid water content in precipitation shafts, one further more extreme case has been included. Table 3.6 shows the result of scaling the Rain 50 drop concentrations by factors of 2, 3, 4 and 5, corresponding to liquid water contents of 4.2, 6.3, 8.4, and 10.5 g m^{-3}, respectively. Cloud thickness is once again 9 km with base height of 500 m and solar zenith angle of $\theta = 0°$. These cases, while extreme, place an upper limit on cloud absorptance values.

From Tables 3.5 and 3.6 we can see that doubling the concentration of large particles significantly increases cloud reflectance from 44.5% to 51.1%. However, cloud absorptance is increased only slightly from 39% to 40.5%. Further increases in particle concentration have only insignificant effects upon the bulk radiative characteristics. Increasing the drop concentration to $5 * 10^{-3}$ cm^{-3} increases cloud reflectance to 55.7% while increasing cloud absorptance to only 41%.

Increasing the concentration of drops in the Rain 10 distribution by an order of magnitude (10^{-2} cm^{-3}) provides a liquid water content of 5.1 g m^{-3}, a value intermediate between the Rain 50∗2 and Rain 50∗3 cases. For such a large concentration of particles cloud reflectance increases to 57.5% while cloud absorptance increases to 38%. Therefore, it appears that cloud reflectance in thick clouds is primarily a function of

TABLE 3.6. Percent cloud reflectance *(R)*, transmittance *(T)* and absorptance *(A)* for the Rain 50 drop size distribution with drop concentrations scaled by factors of 2, 3, 4 and 5 as shown. Cloud thickness is 9 km with cloud top at 9.5 km. Solar zenith angle $\theta = 0°$.

Drop size distribution		←			Wavelength region (μm)				→		
		0.55	0.76	0.95	1.15	1.4	1.8	2.8	3.3	6.3	Total
Rain 50∗2	*R*	82.3	74.2	41.0	22.2	1.6	0.6	0.1	0.4	0.1	51.1
	T	17.3	8.3	0.7	0.1	0.0	0.0	0.0	0.0	0.0	8.4
	A	0.4	17.5	58.3	77.7	98.4	99.4	99.9	99.6	99.9	40.5
Rain 50∗3	*R*	87.4	75.1	41.9	22.9	1.6	0.6	0.2	0.4	0.1	53.6
	T	12.0	3.5	0.1	0.0	0.0	0.0	0.0	0.0	0.0	5.5
	A	0.6	21.4	58.0	77.1	98.4	99.4	99.8	99.6	99.9	40.9
Rain 50∗4	*R*	90.0	75.3	42.4	23.3	1.7	0.6	0.2	0.4	0.1	54.9
	T	9.2	1.5	0.1	0.0	0.0	0.0	0.0	0.0	0.0	4.1
	A	0.8	23.2	57.5	76.7	98.3	99.4	99.8	99.6	99.9	41.0
Rain 50∗5	*R*	91.7	75.3	42.8	23.6	1.7	0.6	0.2	0.4	0.1	55.7
	T	7.4	0.7	0.0	0.0	0.0	0.0	0.0	0.0	0.0	3.3
	A	0.9	24.0	57.2	76.4	98.3	99.4	99.8	99.6	99.9	41.0

drop concentration, while cloud absorptance is a function of drop size as well as drop concentration. Even for comparable values of liquid water content the larger drop mode provides larger values of absorption.

The results in Tables 3.5 and 3.6, extreme as they may appear, may not actually put a limit on cloud absorptance. Consider the Rain L′, Rain 10 and Rain 50 distributions shown in Table 3.5. The large increase in cloud absorptance between Rain L′ (30%) and Rain 50 (39%) was due to the larger drop sizes present in the Rain 50 distribution. However, for extreme rainfall conditions the drop spectrum can be concentrated at even larger drop sizes. Rainfall rates of 150 to 250 mm h^{-1} are reported (Semplak, 1970). In such cases one might expect drop distributions with an effective radius of 1 mm or larger. Since the results in Table 3.5 suggest that drop size is extremely important at the shorter wavelengths, one might expect even larger absorptance for larger drop sizes in the 0.7–1.3 μm spectral regions. However, to extend these results to such large size distributions is beyond the scope of the present work. Nevertheless, Welch and Cox (1978) have shown that in the microwave region the Marshall and Palmer (1947) drop size distribution function gives larger attenuation values than do the Deirmendjian drop size distributions, even for equivalent rainfall rates. Therefore, the question of the effect of the large drop size distribution function upon cloud radiative characteristics, at least for extremely large-particle sizes, remains open.

While the results in the above tables show impressive values of cloud absorptance values for a solar zenith angle of $\theta = 0°$, the measurements by Reynolds *et al.* were often taken at larger solar zenith angles. Table 3.7 contains bulk cloud radiative characteristics similar to those given in Table 3.5 for a solar zenith angle of 60°.

As expected, increasing the solar zenith angle increases cloud reflectance. The variations in cloud reflectance appear to be largest for those cases in which transmittance is largest. Increasing the solar zenith angle from $\theta = 0°$ to 60° doubles the cloud optical depth.

Cloud absorptance in this very thick cloud was surprisingly unaffected by such a large variation in solar zenith angle. The small-particle C.5 distribution decreased in absorptance from 21% to about 17% with this increase in solar zenith angle. However, the Rain L distribution increased its absorptance from 20 to 24%. This effect was undoubtedly due to the fact that the Rain L distribution has a small optical depth; increasing solar zenith angle increased optical depth and the scattering field which then increased droplet absorption. However, in a well-developed scattering field, such as with the C.5 or Rain 50 distributions, increased solar zenith angle merely scatters additional radiation from the cloud at the expense of absorption.

Increasing the concentration of drops at large zenith angles increases cloud reflectance and leaves absorptance relatively unchanged. Increasing the concentration of drops in the C.6 distribution increases reflectance to 67.7%; a similar increase for the Rain 10 distribution leads to a value of 63.4%.

It should be noted that real clouds are not homoge-

TABLE 3.7. Percent cloud reflectance (*R*), transmittance (*T*) and absorptance (*A*) for the C.5, C.6, Rain L, Rain L′, Rain 10 and Rain 50 drop size distributions as a function of spectral region. Cloud thickness is 9 km with cloud top at 9.5 km. Solar zenith angle $\theta = 60°$.

Drop size distribution		Wavelength region (μm)									
		0.55	0.76	0.95	1.15	1.4	1.8	2.8	3.3	6.3	Total
C.5	*R*	97.9	96.9	87.6	82.6	53.1	39.6	0.7	1.3	1.8	82.2
	T	2.1	1.6	0.2	0.0	0.0	0.0	0.0	0.0	0.0	1.1
	A	0.02	1.5	12.2	17.4	46.9	60.4	99.3	98.7	98.2	16.7
C.6	*R*	49.1	49.0	39.9	33.9	15.1	9.8	0.2	0.7	0.2	37.8
	T	50.9	50.8	39.2	31.4	11.9	7.1	0.1	0.1	0.0	37.8
	A	0.01	0.2	20.9	34.7	73.0	83.1	99.7	99.2	99.8	24.4
Rain L	*R*	23.7	24.0	17.7	12.2	4.0	1.9	0.2	0.4	0.1	17.0
	T	76.2	75.9	60.0	50.2	22.4	18.8	5.0	8.0	3.7	58.4
	A	0.1	0.1	22.3	37.6	73.6	79.3	94.8	91.6	96.2	24.6
Rain L′	*R*	66.7	56.8	52.3	39.6	8.3	2.5	0.2	0.7	0.2	48.2
	T	22.7	32.2	24.0	18.8	0.8	0.1	0.0	0.1	0.0	22.8
	A	0.6	1.0	23.7	41.6	90.9	97.4	99.8	99.3	99.8	29.0
Rain 10	*R*	64.2	64.3	49.4	35.6	4.3	1.6	0.3	0.8	0.2	45.5
	T	35.8	33.5	21.3	13.9	0.2	0.1	0.0	0.0	0.0	23.3
	A	0.04	2.2	29.3	50.5	95.5	98.3	99.7	99.2	99.8	31.2
Rain 50	*R*	79.2	77.6	51.9	33.7	4.2	1.7	0.4	1.2	0.2	53.4
	T	20.7	14.6	4.5	1.3	0.0	0.0	0.0	0.0	0.0	11.2
	A	0.1	7.8	43.6	65.0	95.8	98.3	99.6	98.8	99.8	35.4

neous neither vertically nor horizontally. While Chapter 2 demonstrated that vertical variations in liquid water content do not significantly affect bulk radiative cloud characteristics, it has not been demonstrated that horizontal variations may also be neglected. This topic is discussed further in Chapter 6.

3.2.3 BIMODAL DROP SIZE DISTRIBUTIONS IN THIN CLOUDS

The cloud models considered above have concentrated upon monomodal large drop size distributions. However, there is a range of drop sizes in real clouds. The results of the previous sections show that the scattering field is less developed for large drops than for small drops. However, for bimodal size distributions, the scattering field and cloud reflectance are primarily determined from the small droplet size distribution. In precipitation shafts Sartor and Cannon (1977) reported that the small droplet concentration is depleted, often in as little as 60 s, to about 1% of its former value. However, such precipitation shafts are surrounded by the cloud body composed primarily of small droplets. Therefore, even within precipitation shafts, one would expect a very pronounced scattering field, due not to the small particles within the precipitation shaft but to the droplets surrounding this region.

Table 3.8 shows cloud radiative characteristics for a variety of bimodal distributions. In each case the C.5 distribution, representing the small drops, is combined with each of the larger drop distributions to form the bimodal distribution. Cloud base height is 500 m with cloud thickness varying from 500 m to 3 km. A second set of results are also shown for a cloud base height of 5 km.

TABLE 3.8. Percent cloud reflectance *(R)*, transmittance *(T)* and absorptance *(A)* for various bimodal size distributions. A second set of calculations are shown in which the droplet C.5 concentration has been decreased by an order of magnitude. Cloud thicknesses range from 500 to 3000 m for cloud base heights of 500 and 5000 m. Solar zenith angle $\theta = 0°$.

Drop size distributions		Cloud base (m) 500	500	500	5000	5000	5000
		Cloud thickness (m) 500	1000	3000	500	1000	3000
C.5 + C.6	*R*	59.3	71.4	79.8	57.4	68.4	76.1
	T	33.1	18.4	6.0	31.3	17.0	5.5
	A	7.6	10.2	14.2	11.3	14.6	18.4
C.5 + Rain L	*R*	59.0	71.2	79.7	57.1	68.2	76.0
	T	33.4	18.6	6.1	31.6	17.2	5.6
	A	7.6	10.2	14.2	11.3	14.6	18.4
C.5 + Rain L′	*R*	59.3	71.1	79.1	57.1	67.8	75.1
	T	32.4	18.0	5.7	30.5	16.6	5.3
	A	8.3	10.9	15.2	12.4	15.6	19.6
C.5 + Rain 10	*R*	58.9	70.6	78.5	56.6	67.2	74.4
	T	32.4	18.0	5.7	30.4	16.5	5.1
	A	8.7	11.4	15.8	13.0	16.3	20.5
C.5 + Rain 50	*R*	58.6	69.5	76.1	55.6	65.3	71.3
	T	30.4	16.6	5.0	28.2	15.0	4.4
	A	11.0	13.9	18.9	16.2	19.7	24.3
C.5/10 + C.6	*R*	13.2	23.1	46.2	13.0	22.5	44.6
	T	83.2	70.0	38.7	81.6	67.8	36.9
	A	3.6	6.9	15.1	5.4	9.7	18.5
C.5/10 + Rain L	*R*	12.0	20.8	43.4	11.8	20.3	41.9
	T	84.6	72.6	41.8	83.1	70.5	40.0
	A	3.4	6.6	14.8	5.1	9.2	18.1
C.5/10 + Rain L′	*R*	14.4	25.1	47.8	14.0	24.1	45.4
	T	80.7	65.9	34.4	78.7	63.1	32.2
	A	4.9	9.0	17.8	7.3	12.8	22.4
C.5/10 + Rain 10	*R*	13.9	24.1	46.3	13.5	23.0	43.7
	T	80.7	66.0	34.6	78.4	62.9	32.1
	A	5.4	9.9	19.1	8.1	14.1	24.2
C.5/10 + Rain 50	*R*	17.6	29.9	50.4	16.7	27.9	46.6
	T	73.4	55.8	25.8	69.6	51.8	24.5
	A	9.0	14.3	23.8	13.7	20.3	29.9

The bimodal combinations of C.5 and C.6 as well as C.5 and Rain L provide bulk radiative characteristics similar to the monomodal C.5 distribution. Reflectance increases from 59 to 80% as cloud thickness increases from 500 m to 3 km; the corresponding increase in cloud absorptance is from 7.6 to 14%. Increasing cloud base height from 500 m to 5 km decreases cloud reflectance and increases cloud absorptance. The value of absorptance varies between 11 and 18.4% for cloud thickness ranging between 500 m and 3 km.

Increasing the Rain L drop concentration an order of magnitude (Rain L′) leaves cloud reflectance unchanged while increasing absorptance by about 1%. For the larger drop distributions (Rain 10 and Rain 50), the value of cloud reflectance is decreased slightly while that of cloud absorptance is increased. For the Rain 50 drop size distribution cloud absorptance for a cloud base height of 500 m increases from 11 to 19% as cloud thickness increases from 500 m to 3 km. These large values of cloud absorptance are greater than for either the monomodal C.5 or monomodal Rain 50 distributions in thin clouds. Note that for a cloud thickness of 3 km the monomodal Rain 50 distribution provides a larger cloud absorptance than does the bimodal distribution. This once again illustrates that cloud absorptance is a function of both the scattering field as well as the droplet single-scattering albedo.

Changing cloud base height from 500 m to 5 km significantly increases cloud absorptance, with values ranging from about 16 to 24% as cloud thickness increases from 500 m to 3 km. For a cloud thickness of 500 m both the monomodal C.5 and the monomodal Rain 50 distributions provide cloud absorptance values of 11% (see Table 3.4); the bimodal combination, however, provides a value of 16.2%. For a cloud 3 km thick, the monomodal C.5 distribution provides a cloud absorptance value of 18%, while that of the monomodal

Rain 50 provides a value of 30%. The bimodal combination provides a value of 24%. For clouds of small optical depth (≲ 30–40) and bimodal size distributions, the presence of small droplets increases the scattering field and increases absorption. However, for clouds of large optical depth (≳ 100) small droplets tend to scatter more radiation from the cloud, thereby decreasing cloud absorption.

The above results are, of course, highly dependent upon the actual concentration of drops, as well as drop size distribution. Table 3.8 also gives bulk radiative characteristics for these same bimodal drop size distributions for the situation in which the small droplet C.5 concentration is decreased by an order of magnitude (to 10 cm^{-3}).

Most striking is the sharp decrease in cloud reflectance found for all of the bimodal size distributions with the decrease in small droplet concentration. Values of cloud reflectance now range from about 15 to 50% for corresponding cloud thicknesses of 500 m and 3 km. For thin clouds (500 m), cloud reflectance has decreased from about 60 to 13%, a 47% change with the decrease in the C.5 concentration. For thicker clouds (3 km), this decrease was smaller, about 30–35%. These results again show that the small droplets dominate scattering, particularly for thin clouds, and that the large particles have a more dominant effect on radiative characteristics in thicker clouds.

For the thin clouds (≲ 1 km), decreasing the small droplet concentration also decreased the value of cloud absorptance. However, for thick clouds (≳ 3 km) decreasing the C.5 concentration led to increasing values of cloud absorptance. For the C.5 and Rain 50 bimodal size distributions, a decrease in the C.5 concentration by an order of magnitude increased cloud absorptance from 19 to 24% for cloud base height of 500 m; similar increases from 24 to 30% occurred for a cloud base height of 5 km.

For further decreases in C.5 droplet concentration values of cloud absorptance and cloud reflectance approach the monomodal values given in Tables 3.3 and 3.4.

To model Sartor and Cannon's (1977) report of very large drop concentrations and frozen water contents in convectively active clouds the Rain 50 distribution has been scaled upward in drop concentration by factors ranging between 2 and 5 (Table 3.9). Cloud thickness is 3 km, cloud base height is 3 km and solar zenith angle is 0°. Various bimodal size distributions have been assumed.

Using the C.5 small droplet concentration, cloud reflectance decreases steadily and cloud absorptance increases steadily as the concentration of large drops increases. However, even for the extreme case in which liquid water content is 10.5 g m^{-3} (Rain 50*5), cloud absorptance is only 29.4%. An increase in large-particle concentration to Rain 50*10 has about the same effect on cloud absorption as decreasing the small-particle concentration by an order of magnitude (C.5/10 in Table 3.8). Table 3.9 also shows the effect of increasing the large particle concentration while decreasing the small particle concentration. Decreasing the C.5 concentration by an order of magnitude (C.5/10) increases cloud absorptance by about 5%. However, the value of cloud absorptance only increases by about 3% as the liquid water content increases from 4 to about 10 g m^{-3}.

TABLE 3.9. Percent cloud reflectance (R), transmittance (T) and absorptance (A) for various bimodal drop size distributions. The large-drop Rain 50 distribution has been scaled by factors ranging between 2 and 5. The concentration of small drops (C.5 distribution) has been scaled by factors of 0.1 and 0.01; the C.6 distribution has also been used to model the small drops. Cloud thickness is 3 km, cloud base height is 3 km, and solar zenith angle $\theta = 0°$.

Rain drop size distribution		Cloud drop size distribution + C.5	+ C.5/10	+ C.5/100	+ C.6
Rain 50*2	R	69.8	51.2	43.5	44.2
	T	4.2	17.2	24.1	23.4
	A	26.0	31.6	32.4	32.4
Rain 50*3	R	68.7	53.7	48.7	49.2
	T	3.8	13.1	17.2	16.8
	A	27.5	33.2	34.1	34.0
Rain 50*4	R	67.9	55.3	51.9	52.1
	T	3.5	10.7	13.1	13.0
	A	28.6	34.0	35.0	34.9
Rain 50*5	R	67.3	56.6	53.9	54.1
	T	3.3	8.9	10.7	10.5
	A	29.4	34.5	35.4	35.4

Decreasing the concentration of small drops (C.5/100) by another order of magnitude (to 1 cm^{-3}) leads to a further decrease in cloud reflectance and a corresponding increase in cloud absorptance of about 1%. At this point the large drops dominate the cloud radiation characteristics, even though the small droplet concentration is three orders of magnitude larger than the large droplet concentration. Cloud reflectance increases with the increase in large drop concentration.

As a final example we replace the C.5 drop size distribution with the C.6 distribution. This bimodal size distribution provides similar results to that obtained using the C.5/100 distribution. The C.6 distribution also has a drop concentration of 1 cm^{-3}. It appears that droplet concentrations of about 1 cm^{-3}, independent of size distribution, have negligible effects upon the cloud radiative characteristics in the presence of large drops with concentrations of 10^{-3} cm^{-3}; this is true for clouds of moderate thickness. Therefore, these results, when combined with the concentration data reported by Sartor and Cannon, indicate that within precipitation shafts the presence of small droplets may be essentially neglected in the determination of the radiative characteristics of these clouds.

3.2.4 Bimodal drop size distributions in thick clouds

The results discussed in the previous section for bimodal drop size distributions have been for thin and moderately thick clouds. To complete this picture we must also examine results for thick clouds. Table 3.10 shows calculations as a function of spectral region for a number of bimodal size distributions. Cloud thickness is 9 km with cloud base height of 500 m and solar zenith angle $\theta = 0°$.

Table 3.5 gave the contributions of the individual monomodal drop size distributions to the radiation characteristics for such a thick cloud. In particular the C.5 small-particle distribution function produces a reflectance value of 77.5% with an absorptance value of 20.9%. The values of reflectance given for the bimodal size distributions are similar to that of the C.5 distribution alone. This once again demonstrates the fact that the scattering field in a bimodal distribution is determined primarily by the large concentration of small particles. However, cloud reflectance is substantially reduced in the C.5 + Rain 50 distribution due to the strong absorption by the large particles.

Of particular interest is the fact that the bimodal size distributions do not produce large values of cloud absorptance. Even for the C.5 + Rain 50 bimodal size distribution absorptance is only about 27%. This value is greater than that due to the small particles alone (C.5 in Table 3.5, with a value of about 21%). The large drops alone (Rain 50 in Table 3.5) produce a much larger absorptance value (39%). Therefore, it appears that the presence of large concentrations of small droplets, while increasing the scattering field, decreases total cloud absorption.

In order to examine this behavior over the various spectral regions, comparison of Tables 3.5 and 3.10 is necessary. In the spectral region with wavelengths < 0.7 μm (labeled 0.55 μm), the C.5 distribution provides an absorptance value of 0.04%; the Rain 50 distribution gives a value more than double this amount, or 0.1%. The bimodal distribution gives an absorptance value of 0.2%. Therefore, within this spectral region increased scattering leads to increased absorption. However, this value of absorptance, even assuming this large bimodal distribution, is so small as to be negligible for most practical considerations.

In the 0.7–0.8 μm region the C.5 distribution gives an absorptance value of 2% while the Rain 50 distribution produces a value of 10% and the combined bimodal distribution produces 8%. This result indicates that the small particle scattering reduces cloud absorption in this region. However, the Rain L′ distribution increases this value to about 3%. Likewise, the Rain 10 distribution produces an absorptance value of 2.4% which is increased to about 4% for the C.5 + Rain 10 distribution. As a general rule then, we may conclude that for wavelengths below 0.8 μm increased scattering due to the presence of large concentrations of small drops usually increases cloud absorptance. However, just the opposite behavior occurs within the water vapor bands.

Within the first four water vapor bands (labeled 0.95, 1.15, 1.4 and 1.8 μm), the large particles produce significantly larger values of absorption than do the small particles (C.5). Including the C.5 distribution

Table 3.10. Percent cloud reflectance *(R)*, transmittance *(T)* and absorptance *(A)* for five bimodal drop size distributions as a function of spectral region. Cloud thickness is 9 km with cloud top at 9.5 km. Solar zenith angle $\theta = 0°$.

Bimodal drop size distribution		Wavelength region (μm)									
		0.55	0.76	0.95	1.15	1.4	1.8	2.8	3.3	6.3	Total
C.5 + C.6	*R*	96.9	95.5	82.6	75.7	39.6	25.5	0.2	0.4	0.6	77.3
	T	3.0	2.2	0.3	0.0	0.0	0.0	0.0	0.0	0.0	1.6
	A	0.1	2.3	17.1	24.3	60.4	74.5	99.8	99.6	99.4	21.1
C.5 + Rain L	*R*	96.9	95.4	82.6	75.6	39.6	25.7	0.2	0.4	0.6	77.3
	T	3.0	2.3	0.3	0.0	0.0	0.0	0.0	0.0	0.0	1.6
	A	0.1	2.3	17.1	24.4	60.4	74.3	99.8	99.6	99.4	21.1
C.5 + Rain L′	*R*	96.5	95.0	81.4	73.9	35.9	22.5	0.2	0.4	0.6	76.1
	T	2.7	1.9	0.1	0.0	0.0	0.0	0.0	0.0	0.0	1.3
	A	0.8	3.1	18.5	26.1	64.1	77.5	99.8	99.6	99.4	22.6
C.5 + Rain 10	*R*	97.0	94.3	79.7	71.8	33.4	21.6	0.2	0.4	0.6	75.4
	T	2.9	1.6	0.1	0.0	0.0	0.0	0.0	0.0	0.0	1.4
	A	0.1	4.1	20.2	28.2	66.6	78.4	99.8	99.6	99.4	23.2
C.5 + Rain 50	*R*	97.0	91.3	73.8	63.6	24.0	16.1	0.2	0.4	0.6	71.9
	T	2.8	0.7	0.0	0.0	0.0	0.0	0.0	0.0	0.0	1.3
	A	0.2	8.0	26.2	36.4	76.0	83.9	99.8	99.6	99.4	26.8

to produce a bimodal drop size spectrum, strongly reduces cloud absorptance and strongly increases cloud reflectance within these same water vapor bands. In the last three water vapor bands (labeled 2.8, 3.3 and 6.3 μm), absorption is nearly complete, independent of monomodal or bimodal drop size distribution. We may conclude from these calculations that the potentially strong effect of large droplets in thick clouds is constrained to the region of the first four water vapor bands ($\sim$ 0.85–2.3 μm).

In Chapter 2 it was shown that the various small monomodal size distributions produced similar values of cloud absorptance but substantially different values of reflectance. Table 3.11 explores this topic more fully. Six small-particle drop size distributions are considered; their radiative characteristics defined in Chapter 2. Cloud thickness is 9 km with cloud base height at 500 m, and solar zenith angle $\theta = 0°$. Column A shows values of cloud reflectance, transmittance and absorptance for the various monomodal, small-particle drop size distributions. For this rather thick cloud all of the distributions give similar values of reflectance (76–78%). However, there is more variability in cloud absorptance, with values ranging from 20.3 to 23.3%. Note that the size distributions representing cloud bases give smaller values of absorptance than do the size distributions representing cloud tops. The size distributions representing cloud tops are less sharply peaked (Fig. 2.1) and include larger particle sizes.

TABLE 3.11. Percent cloud reflectance *(R)*, transmittance *(T)* and absorptance *(A)* for six small-particle size distributions (defined in Chapter 2). Values given are for the entire solar spectrum. Column A gives the values for the monomodal size distribution; column B values of the bimodal distribution which is defined as the combination of small-particle distribution with the Rain 50 distribution; column C values for the bimodal distribution with the small-particle concentration decreased to 10% of its original value; and column D the same values for the small-particle concentration decreased to 1% of its original value. Cloud thickness is 9 km with cloud top at 9.5 km. Solar zenith angle $\theta = 0°$.

Small-particle size distribution		A Mono-modal	B Bimodal combined with Rain 50	C Bimodal Small/10	D Bimodal Small/100
Nimbostratus top	*R*	76.0	73.1	61.0	48.3
	T	0.7	0.6	4.8	13.3
	A	23.3	26.3	34.2	38.4
Nimbostratus base	*R*	77.6	70.3	55.1	46.2
	T	1.8	1.5	7.4	15.0
	A	20.6	28.2	36.5	38.8
Stratus top	*R*	77.6	71.6	56.8	46.7
	T	1.3	1.1	7.3	14.6
	A	21.1	27.3	35.9	38.7
Stratus base	*R*	76.7	67.4	52.1	45.5
	T	3.0	2.4	10.4	15.6
	A	20.3	30.2	37.5	38.9
Stratocumulus top	*R*	77.8	73.6	60.4	47.9
	T	9.7	0.7	5.2	13.6
	A	21.5	25.7	34.4	38.5
Stratocumulus base	*R*	77.0	68.5	53.0	45.7
	T	2.6	2.1	9.8	15.4
	A	20.4	29.4	37.2	38.9

Column B in Table 3.11 shows the radiative characteristics resulting from a bimodal drop size distribution made up of a combination of each of the small-particle distributions together with the Rain 50 distribution. The bimodal distributions result in decreased values of cloud reflectance and substantially larger values of cloud absorptance. However, values of both cloud reflectance and cloud absorptance now vary strongly as a function of the small-particle drop size distribution. Note that the bimodal drop size distributions incorporating the small-particle distributions at cloud bases have smaller values of cloud reflectance and larger values of cloud absorptance than do the distributions representing droplets at cloud tops. Cloud reflectance values range between 67.4 and 73.6%, while cloud absorptance values range between 25.7 and 30.2%. In any case, these values of cloud absorptance are significantly larger than the maximum of 20% reported by Twomey (1976).

The reason for this variation in radiative characteristics may be understood by referring to Tables 2.1 and 2.2. There we find that the stratus base size distribution has the smallest values of attenuation parameters as well as the smallest liquid water content; yet this distribution has the largest value of absorptance in Table 3.11, and the smallest value of reflectance. The bimodal combination of stratus base with Rain 50 provides smaller bimodal attenuation coefficients than the combination of nimbostratus top with Rain 50, with less scattering (and reflectance). The bimodal single-scattering albedo is smallest for the bimodal distribution with the smallest attenuation coefficients, leading to increased absorption.

Comparison between the last 3 size distributions in Tables 3.5 and 3.10 shows that for thick clouds, bimodal size distributions produce smaller cloud absorptance values than do large-particle distributions alone. Columns C and D of Table 3.11 depict conditions typical of a precipitation shaft i.e. a reduced small-droplet population in the presence of larger precipitating droplets. Column C considers the same six bimodal size distributions discussed previously, but with the small-particle concentrations decreased by an order of magnitude (to 10^1 cm^{-3}). Column D considers the same situation, but with the small-particle concentration reduced by an additional order of magnitude (to 1 cm^{-3}).

Decreasing the small-particle concentrations by an order of magnitude (Column C) results in significantly smaller values of reflectance and correspondingly larger values of both absorptance and transmittance. Values of cloud reflectance for these bimodal distributions range between 52 and 61%, with the largest values of reflectance correlated with the largest liquid water contents contained in the small-particle mode. These large differences in cloud reflectance values exist even though equal numbers of particles (10^1 cm^{-3}) are modeled in all six cases. Such behavior may make remote sensing of clouds in the solar spectrum more difficult than previously imagined.

Note that decreasing the small-particle concentration another order of magnitude (to 1 cm^{-3}) produces results similar to that of the large-particle monomodal size distribution (Table 3.5). In this case, values of cloud reflectance and absorptance are almost independent of the small-particle size distribution. The concentration of small particles is three orders of magnitude less than large particles which now totally dominate the radiation field. The results in Column D simulate the conditions reported by Sartor and Cannon (1977) in which the small-particle concentration is reduced to about 1% of its normal value within a precipitation shaft. In such cases the radiation field may be determined from the characteristics of the large particles alone, neglecting the small particles altogether.

Cloud reflectance is relatively unaffected by the choice of the small-particle size distribution function in two instances: 1) when the large particles dominate (Column D) and 2) when the small particles dominate (Column B) in thick clouds. Only in the intermediate range of small-particle concentrations is there a significant dependence of cloud reflectance upon the specific small droplet size distribution. Since small-particle size distributions with such concentrations are often measured in clouds, variations in the small droplet as well as large drop size distributions may lead to very large variations in cloud reflectance values as measured by remote sensing techniques.

For any of the small-particle size distributions, variations in droplet concentration (comparison of Columns B, C and D in Table 3.11) may lead to extremely large variations in measured values of cloud reflectance. Variations in reflectance of this magnitude may also occur quite rapidly, since Sartor and Cannon report that the small droplets may be depleted within precipitation shafts in convectively active clouds in as little as 60 s. Obviously the possibility of such wide variations in such short periods of time may place severe restrictions upon remote sensing applications within the solar spectrum.

These same considerations make the anomalous results reported by Reynolds *et al.* more plausible. Comparison of Columns B, C and D shows that as the concentration of small drops decreases to about 1 cm^{-3}, there is a corresponding increase in the value of cloud absorptance. Furthermore, the large values of cloud absorptance (38%) calculated in Table 3.11 are for liquid water contents of only 2 g m^{-3}. Considering the values reported by Sartor and Cannon along with calculations shown in Table 3.6, one is led to the conclusion that the extremely large values of cloud absorptance reported by Reynolds *et al.* may in fact be real.

3.2.5 Heating rates

Results in previous sections have dealt with bulk radiative characteristics such as cloud reflectance and absorptance. Another quantity of significance, particularly to dynamical studies, is the radiative heating rate. The effect of monomodal drop size distributions, cloud thickness, cloud base height and solar zenith angle upon cloud-averaged heating rates were discussed in Chapter 2.

Cloud-averaged heating rates for the various bimodal drop size distributions shown in Table 3.8 are given in Table 3.12. The C.5 + C.6 and C.5 + Rain L distributions give identical results, the absorptance being dominated by the small droplets. Even though cloud absorptance increases with increasing cloud thickness, cloud-averaged heating rates decrease. The reason for this behavior is that cloud depth increases more rapidly than does absorbed radiation. Consider the C.5 + C.6 size distribution. A cloud 3 km in thickness absorbs twice as much energy as a cloud 500 m in thickness; however, the 3 km cloud is six times thicker. Therefore, the cloud-averaged heating rate for the large cloud is about one-third as large as that for the small one. For a 500 m thick cloud, cloud-averaged heating rates are about three times larger for a cloud top at 5.5 km than at 1 km, even though the difference in absorptance is much smaller.

Heating rates for most of the bimodal size distributions in combination with the C.5 distribution have similar values. The C.5 + Rain 50 bimodal size distribution provides somewhat larger values. In this case, heating rates of 0.75°C h^{-1} are calculated for a 500 m

TABLE 3.12. Cloud-averaged heating rates (°C h^{-1}) for the bimodal drop size distributions, cloud thicknesses and cloud base heights shown in Table 3.8. Solar zenith angle $\theta = 0°$.

Drop size distribution	Base height (m) 500 / Thickness (m) 500	500 / 1000	500 / 3000	5000 / 500	5000 / 1000	5000 / 3000
C.5 + C.6	0.52	0.36	0.20	1.45	0.98	0.48
C.5 + Rain L	0.52	0.36	0.20	1.45	0.98	0.48
C.5 + Rain L′	0.57	0.39	0.21	1.59	1.05	0.52
C.5 + Rain 10	0.60	0.41	0.22	1.67	1.10	0.54
C.5 + Rain 50	0.75	0.50	0.27	2.07	1.32	0.64
C.5/10 + C.6	0.23	0.24	0.21	0.69	0.65	0.49
C.5/10 + Rain L	0.23	0.24	0.21	0.69	0.65	0.49
C.5/10 + Rain L′	0.33	0.32	0.25	0.93	0.86	0.59
C.5/10 + Rain 10	0.37	0.35	0.27	1.04	0.95	0.64
C.5/10 + Rain 50	0.62	0.51	0.33	1.72	1.37	0.79

thick cloud with base height of 500 m, a value about 50% greater than that found with the monomodal C.5 distribution. For a 3 km thick cloud, however, the differences are smaller (0.27°C h^{-1} compared to 0.20°C h^{-1}).

Table 3.12 shows that decreasing the concentration of small droplets by an order of magnitude has a very pronounced effect on the cloud-averaged heating rates; heating rates decrease by a factor of 2 for thin clouds (500 m) for the C.6 and Rain L large-particle modes. Increasing the concentration of large particles by an order of magnitude (Rain L′) now strongly increases the value of the bimodal heating rate. The decrease of the small-particle concentration in thick clouds (3000 m) leads to an increase in heating rates. In all of these cases there is a direct correlation between cloud absorptance, cloud thickness and cloud-averaged heating rate.

For a cloud 9 km in thickness with a cloud absorptance value of 30%, the solar average heating rate is 0.23°C h^{-1}. Heating rates of 0.03, 0.09, 0.05 and 0.02°C h^{-1} are contributed by the 0.95, 1.15, 1.4, 1.8 and 2.8 μm region bands, with minor contributions at larger and smaller wavelengths; these results are for the C.5/100 + Rain L′ bimodal size distribution. Of course, with larger drop sizes, there are increased heating rates at the smaller wavelengths.

For clouds of uniform optical densities, very large heating rates are found near cloud top (Chapter 2). However, real clouds cannot sustain such large values, since cloud tops would rapidly "burn off". Therefore, it seems plausible that clouds with variations in cloud optical density are to be expected in order to more evenly disperse radiative heating throughout the cloud. Oliver *et al.* (1978) argue that infrared cooling is localized near cloud top, while solar heating is absorbed deeper within the cloud interior. The strong cloud top cooling provides cloud instability and turbulence production to transport the heat absorbed within the cloud body. For fog and low level stratus, their model indicates that cloud tops are not "burned away" by absorption of solar radiation at cloud top, but evaporated away by turbulent transfer of solar radiation absorbed in deeper regions of the cloud body.

3.2.6 Multimodal drop size distributions

The results of Jonas and Mason (1974) suggest that the drop size distribution is relatively flat for the large particle spectrum (Fig. 3.1). However, the drop size distributions considered above consist primarily of a single small-particle mode along with an independent large-particle mode. Fig. 3.1 indicates that the large-particle concentration may grow along with convective cloud activity. To model this case, various combinations of the C.5 and C.6 particle modes are used to represent early cloud development. Combinations of the large-particle Rain L′, Rain 10 and Rain 50 distributions are added in order to provide a more uniform drop spectrum. In this way one may examine the variation in radiative characteristics as progressively larger drop sizes are developed. The results of this study are shown in Table 3.13.

However, we first consider results for the C.5 + C.6 small-particle combination. For a cloud of thickness of 1 km, Table 3.8 showed that the C.5 + C.6 size distribution gives a value of 71.4% for cloud reflectance and a value of 10.2% for cloud absorptance. Slightly smaller values of reflectivity and slightly larger values of absorptance are obtained with the C.5 + Rain L′ and C.5 + Rain 10 bimodal distributions. Table 3.10 gave comparable calculations for a 9 km thick cloud. Values of cloud reflectance are 77.3, 76.1 and 75.4% for the C.5 + C.6, C.5 + Rain L′ and C.5 + Rain 10 bimodal size distributions, respectively; values of absorptance are 21.1, 22.6 and 23.2%, respectively.

Calculations in Table 3.13 demonstrate that the trimodal distributions C.5 + C.6 + Rain L′, C.5 + C.6 + Rain 10 and C.5 + C.6 + Rain 50 provide nearly identical values of bulk radiative characteristics as do the bimodal C.5 + Rain L′, C.5 + Rain 10 and C.5 + Rain 50 distributions. Therefore, the effect of the C.6 distribution is negligible.

The effect of including the Rain L′ distribution with the Rain 10 and Rain 50 modes is to increase cloud absorptance about 1%, from 23.2 to 24.4% and from 26.8 to 27.5%, respectively, in a 9 km thick cloud. Likewise, the inclusion of the Rain 10 mode with the Rain 50 mode increases cloud absorptance by about 1% for this 9 km thick cloud. Combining the Rain L′ + Rain 10 + Rain 50 modes increases cloud absorptance by about 0.5%, to 28.2%, for this thick cloud. Smaller increases are found for the 1 km cloud. The value of cloud reflectance is nearly identical for clouds 1 and 9 km thick. However, absorptance in the thick cloud is about double that in the thin cloud. Cloud heating rates vary from 0.17 to 0.22°C h^{-1} for the 9 km thick cloud, with values of 0.39 to 0.53°C h^{-1} in the 1 km thick cloud.

The variations in cloud radiative characteristics, obtained by including a more complete large-particle size distribution (Table 3.13), are smaller than those variations obtained by the different small-particle distributions (Table 3.11). The value of cloud absorptance is dominated by the presence of large drops. The presence of drops of intermediate size appears to have negligible consequences on the bulk radiative characteristics of clouds.

Table 3.13 also shows calculations for a zenith angle of $\theta = 60°$. For a cloud of 1 km thickness, an

TABLE 3.13. Reflectance R and absorptance A (percent) and cloud averaged heating rates $\partial\theta/\partial t$ (°C h^{-1}) for various multimodal drop size distributions. The left-hand column describes the various large-particle combinations used here. Combinations of the C.5 and C.6 drop size distributions have been used to represent the small-particle modes. C.5/10 and C.5/100 refer to a decrease by one and two orders of magnitude in the C.5 particle concentration; C.6 * 10 refers to an increase by an order of magnitude in the C.6 concentration. Cloud thicknesses are 1 and 9 km with a cloud base height of 500 m. Solar zenith angles are $\theta = 0°$ and 60°.

Large-particle size distribution		Small-particle size distributions															
		C.5 + C.6				C.5/10 + C.6				C.5/10 + C.6 * 10				C.5/100 + C.6 * 10			
		Thickness = 9 km		Thickness = 1 km		Thickness = 9 km		Thickness = 1 km		Thickness = 9 km		Thickness = 1 km		Thickness = 9 km		Thickness = 1 km	
		$\theta = 0°$	$\theta = 60°$	$\theta = 0°$	$\theta = 60°$	$\theta = 0°$	$\theta = 60°$	$\theta = 0°$	$\theta = 60°$	$\theta = 0°$	$\theta = 60°$	$\theta = 0°$	$\theta = 60°$	$\theta = 0°$	$\theta = 60°$	$\theta = 0°$	$\theta = 60°$
Rain L′	R	76.0	80.9	71.2	80.6	60.9	68.6	27.5	48.3	66.6	72.8	43.3	60.8	63.0	69.9	35.0	54.1
	A	22.6	18.1	11.0	6.8	28.3	24.0	9.4	7.3	28.3	23.6	11.7	8.1	29.5	25.0	11.4	8.4
	$\frac{\partial\theta}{\partial t}$	0.17	0.069	0.39	0.12	0.22	0.092	0.34	0.12	0.22	0.090	0.42	0.14	0.23	0.096	0.41	0.14
Rain 10	R	75.4	80.4	70.8	80.3	59.1	66.8	26.6	47.4	65.5	71.8	43.3	60.2	61.8	68.7	34.2	53.4
	A	23.2	18.7	11.5	7.2	30.2	25.8	10.3	8.1	29.3	24.6	12.4	8.7	30.8	26.2	12.0	9.0
	$\frac{\partial\theta}{\partial t}$	0.18	0.071	0.41	0.12	0.23	0.099	0.37	0.14	0.23	0.094	0.44	0.15	0.24	0.10	0.43	0.15
Rain 50	R	71.9	77.4	69.7	79.2	57.3	64.4	31.9	51.0	62.6	68.9	45.6	61.4	59.6	66.3	38.2	55.8
	A	26.8	21.7	14.0	9.2	35.5	30.6	14.5	11.4	33.4	28.3	15.3	11.1	34.9	30.0	15.2	11.5
	$\frac{\partial\theta}{\partial t}$	0.21	0.083	0.50	0.16	0.27	0.12	0.52	0.19	0.26	0.11	0.55	0.19	0.27	0.11	0.54	0.20
Rain L′ Rain 10	R	74.4	79.6	70.8	80.2	60.1	67.3	30.8	50.7	65.2	71.5	45.5	61.7	62.1	68.8	37.6	55.8
	A	24.4	19.6	12.1	7.7	31.5	26.9	11.8	9.1	30.3	25.5	13.1	9.3	31.7	27.0	13.0	9.6
	$\frac{\partial\theta}{\partial t}$	0.19	0.075	0.43	0.13	0.24	0.10	0.42	0.15	0.23	0.097	0.47	0.16	0.24	0.10	0.46	0.16
Rain L′ Rain 50	R	71.4	76.9	69.8	79.3	58.3	65.2	35.4	53.7	62.7	69.0	47.6	62.8	60.2	66.7	41.0	57.8
	A	27.5	22.3	14.3	9.5	35.7	30.7	15.1	11.7	33.8	28.6	15.7	11.4	35.2	30.1	15.7	11.8
	$\frac{\partial\theta}{\partial t}$	0.21	0.085	0.51	0.16	0.27	0.12	0.54	0.20	0.26	0.11	0.56	0.19	0.27	0.12	0.56	0.20
Rain 10 Rain 50	R	71.2	76.7	69.3	78.9	57.0	64.0	33.1	51.8	62.4	68.6	47.1	62.4	59.8	66.3	40.5	57.4
	A	27.6	22.5	14.6	9.7	36.3	31.5	15.3	12.1	34.0	28.9	16.0	11.7	35.4	30.4	16.0	12.1
	$\frac{\partial\theta}{\partial t}$	0.21	0.086	0.52	0.17	0.28	0.12	0.55	0.21	0.26	0.11	0.57	0.20	0.27	0.12	0.57	0.21
Rain L′ Rain 10 Rain 50	R	70.7	76.3	69.7	79.1	58.6	65.3	38.0	55.5	62.5	68.7	48.9	63.7	60.2	66.6	43.0	59.2
	A	28.2	23.0	14.9	10.0	36.2	31.1	15.9	12.2	34.3	29.1	16.3	11.9	35.6	30.5	16.4	12.3
	$\frac{\partial\theta}{\partial t}$	0.22	0.088	0.53	0.17	0.28	0.12	0.57	0.21	0.26	0.11	0.58	0.20	0.27	0.12	0.59	0.21

increase in solar zenith angle from $\theta = 0°$ to $\theta = 60°$ leads to an increase in cloud reflectance by about 10%, from about 70% ($\theta = 0°$) to about 80% ($\theta = 60°$); corresponding decreases in cloud absorptance values of about 5% are calculated for this situation. For a 9 km thick cloud, increasing the solar zenith angle from $\theta = 0°$ to $\theta = 60°$ leads to an increase in the value of cloud reflectance by about 5% as well as a corresponding decrease in the value of cloud absorptance of 5%. The value of cloud average heating rate decreases by about a factor of 3 with this increase in solar zenith angle.

Decreasing the concentration of particles in the C.5 mode (C.5/10) by an order of magnitude (10 cm^{-3}) leads to a decrease in the value of cloud reflectance for all multimodal size distributions, cloud thicknesses and solar zenith angles. A decrease in cloud reflectance of 10–15% occurred with the decrease in the C.5 concentration for the 9 km thick cloud, at both $\theta = 0°$ and $\theta = 60°$. For a 1 km thick cloud the decrease in C.5 concentration led to a decrease in cloud reflectance by 35–40% for $\theta = 0°$ and 25–30% for $\theta = 60°$. For the thick cloud the decrease in C.5 concentration led to an increase of about 6–8% in the value of cloud absorptance, independent of solar zenith angle or size distribution. For the thin cloud the decrease in C.5 concentration led to an increase in the value of cloud absorptance at $\theta = 60°$. For $\theta = 0°$ an increase or decrease in the value of cloud absorptance occurs, depending upon whether or not there are very large (Rain 50) particles in the distribution.

The next small-particle distribution considered in Table 3.13 retains the C.5/10 concentration and increases the concentration of the C.6 particles (C.6*10) by an order of magnitude (to 10 cm^{-3}). In this case both the C.5 and C.6 sized particles exist in equal numbers. The result of increasing the concentration of these intermediate-sized particles is to increase the value of reflectance by 10–15% in the thin (1 km) cloud, and by 3–6% in the thick (9 km) cloud. For the thin cloud, this increase in particle concentration leads to slightly larger values of cloud absorptance; for the thick cloud, slightly smaller values of cloud absorptance result.

The final small-particle distribution shown in Table 3.13 further decreases the concentration of the C.5 mode (C.5/100) by another order of magnitude (to 1 cm^{-3}) while retaining the concentration of the C.6 mode (C.6*10) at 10 cm^{-3}. Note that this situation is exactly opposite to that of the C.5/10 + C.6 distribution where the C.5 mode had a concentration an order of magnitude larger than that of the C.6 mode. Comparison of these two cases shows that similar values of cloud absorptance occur for similar small-particle concentrations, independent of small-particle size distribution. However, the C.5/100 + C.6*10 distribution, with the large-particle sizes, provides slightly larger values of cloud reflectance than does the C.5/10 + C.6 distribution, particularly for the thin cloud.

The combination of results presented in Table 3.13 shows that the radiative characteristics of clouds for a particular cloud thickness and solar zenith angle are primarily influenced by the concentration of small particles and by the largest particles of the large-particle mode.

As a final case, not shown in Table 3.13, a "flat" distribution of drop sizes was constructed with a uniform concentration of drops (10^{-4} cm^{-3} μm^{-1}) for drop sizes between 50 and 500 μm. Calculations using this large-particle distribution, in combination with the various small-particle modes shown in Table 3.13, gave nearly identical values of cloud radiation characteristics as obtained with the Rain L′ + Rain 10 + Rain 50 size distribution. Therefore, it is once again demonstrated that cloud radiative characteristics are somewhat insensitive to the actual distribution of large drops.

3.2.7 Variations in cloud radiative characteristics as a function of bimodal drop size distributions for clouds of constant optical thickness

For monomodal drop size distributions, Section 2.4.5 discussed variations in bulk radiative characteristics as a function of drop size distributions, cloud thickness and drop concentration, for clouds of constant optical thickness. It was shown that clouds of the same optical thickness may have quite different absorptance and reflectance characteristics, particularly in the 0.8–2.5 μm region. Such considerations are of even greater significance for bimodal drop size distributions.

The same conditions as given in Table 2.12 are assumed in Table 3.14; cloud top is at 2500 m with solar zenith angle of $\theta = 0°$. The bimodal size distribution is once again formed by adding the Rain 50 drop size distribution to represent the large drops. This distribution has a drop concentration of $N = 10^{-3}$ cm^{-3} and a value of β_e of about 3.5 km^{-1}. The optical thickness for this bimodal distribution has been maintained at 50 by decreasing the small droplet concentrations appropriately.

The stratus base small particle drop size distribution along with the Rain 50 large particle drop size distribution is considered first. Cloud thickness is 2 km with small-particle drop concentration of 90 cm^{-3}; the large-particle drop concentration is 10^{-3} cm^{-3}. Table 3.14 shows that the presence of the large drops decreases the value of reflectance over that of the

TABLE 3.14. Comparison of radiative properties of clouds with identical optical depths but different bimodal size distributions, geometric thicknesses and small droplet concentrations.

Small-particle size distribution	Cloud thickness (m)	Small droplet concentration (cm^{-3})		← Wavelength region (μm) → 0.55	0.765	0.95	1.15	1.4	1.8	2.8	3.35	6.3	Total	Average cloud heating rate (°C h^{-1})
Stratus base	2000	90	R	80.5	79.7	68.6	56.4	20.6	15.4	0.4	0.6	0.9	68.7	0.34
			T	19.2	17.3	10.2	4.6	0.0	0.0	0.0	0.0	0.0	13.9	
			A	0.3	3.0	21.2	39.0	79.4	84.6	99.6	99.4	99.1	17.4	
Nimbostratus top	2000	21.5	R	79.1	77.9	66.0	53.0	16.4	10.2	0.2	0.3	0.4	66.5	0.36
			T	20.6	18.9	11.6	5.4	0.0	0.0	0.0	0.0	0.0	15.2	
			A	0.3	3.2	22.4	41.6	83.6	89.8	99.8	99.7	99.6	18.3	
Stratus base	500	390	R	80.4	80.7	77.8	72.8	39.3	29.0	0.4	0.6	1.0	73.4	0.93
			T	19.4	18.2	15.1	11.4	0.7	0.1	0.0	0.0	0.0	15.6	
			A	0.2	1.1	7.1	15.8	60.0	70.9	99.6	99.4	99.0	11.0	
Nimbostratus top	500	98	R	79.4	79.2	75.3	68.9	30.5	18.2	0.2	0.3	0.4	70.9	1.07
			T	20.3	19.4	16.1	12.0	0.4	0.0	0.0	0.0	0.0	16.4	
			A	0.3	1.4	8.6	19.1	69.1	81.8	99.8	99.7	99.6	12.7	
Stratus base	250	800	R	80.1	80.7	79.1	76.0	45.8	33.4	0.4	0.6	1.0	74.4	1.61
			T	19.7	18.5	16.4	13.5	1.6	0.2	0.0	0.0	0.0	16.2	
			A	0.2	0.8	4.5	10.5	52.6	66.4	99.6	99.4	99.0	9.4	
Nimbostratus top	250	198	R	79.3	79.3	76.7	72.0	34.8	20.4	0.2	0.3	0.4	71.8	1.94
			T	20.4	19.6	16.6	13.7	0.9	0.1	0.0	0.0	0.0	16.8	
			A	0.25	1.1	6.7	14.3	64.3	79.5	99.8	99.7	99.6	11.4	

monomodal size distribution (Table 2.12a), while leaving the value of transmittance essentially unchanged. This once again demonstrates that transmittance is a function solely of optical thickness rather than of size distribution. The values of absorptance and cloud-averaged heating rate are increased significantly (by about 75%) over the values obtained for the monomodal distribution.

Table 2.12 showed that the nimbostratus top monomodal drop size distribution, for the same cloud thickness (2000 m) and with drop concentration decreased to $N = 25$ cm^{-3}, provided a smaller value of reflectance and larger value of absorptance. This behavior is retained for the bimodal drop size distributions in Table 3.14. Once again the distribution with the largest drop concentration has the largest value of reflectance and smallest value of absorptance.

Differences between the various monomodal and/or bimodal size distributions only occur for wavelengths between 0.8 and 2.5 μm, with the largest differences occurring at the longer wavelengths. However, there are significant differences between the reflectance values of the monomodal and bimodal distributions even at a wavelength of 0.95 μm. Differences in the value of reflectance obtained using various small-particle modes are much smaller than the differences resulting from the presence of the large-particle mode in this 0.8–2.5 μm region.

Table 3.14 also shows results for a cloud 500 m thick with the small-particle concentrations scaled upward to preserve optical thickness at a value of 50. The large-drop concentration has been preserved at 10^{-3} cm^{-3}. The entire effect is again localized to the 0.8–2.5 μm region. The increase in small-particle concentration has increased the value of reflectance while significantly decreasing the values of absorptance and cloud-averaged heating rates for both small-droplet modes. Nevertheless, the values of reflectance in the presence of the large-particle mode are smaller than for the monomodal modes; and the values of absorptance for the bimodal distributions remain larger than for the corresponding monomodal distributions.

As a final example, the cloud thickness is once again halved with the small-particle concentrations doubled. The value of reflectance is once again increased, while that of absorptance is decreased. In fact the values of reflectance, transmittance and absorptance are very similar to those values obtained in some cases in Table 2.12. The effect of the large-particle mode, therefore, is effectively masked, even in the 0.8–2.5 μm wavelength region, in the presence of large concentrations of small drops. This behavior, therefore, makes remote sensing of size distribution using the solar spectrum more difficult. However, note the large difference in reflectance between the 0.765 and 0.95 μm wavelength regions when the small-droplet concentration is low. These large differences may be associated with larger values of cloud absorptance brought about by the presence of large drops. It would be expected that the presence of substantial concentrations of large drops (10^{-4} to 10^{-3} cm^{-3}) occurs at the expense of the small-drop concentrations (Sartor and Cannon, 1977). On the other hand, slight differences between the values of reflectance between the 0.765 and 1.15 μm regions implies the presence of very large concentrations of small drops. The presence or absence of the large-drop mode in this case has essentially no effect upon the values of R, T and A.

Tables 2.12 and 3.14 show that the values of reflectance, and absorptance may change significantly in some cases with variations in monomodal or bimodal drop size distribution, cloud thickness and drop concentration, even though the cloud optical thickness is held constant. However, of perhaps primary importance to cloud dynamics is the variation of absorptance. This value varies by a factor of 2 in the examples used in Tables 2.12 and 3.14. For a cloud of constant optical depth, as determined from the value of reflectance at wavelengths < 0.8 μm, large differences in the values of reflectance at 0.765 and 0.95 μm (and 1.15 μm) are indicative of smaller concentrations of the small-drop mode, thicker clouds and larger values of absorptance. On the other hand, when the difference between the reflectance values for the wavelength regions of 0.765 and 0.95 μm (and 1.15 μm) is small, the cloud has larger concentrations of small droplets and smaller absorptance values. The presence of large-particle concentrations of 10^{-4} to 10^{-3} cm^{-3} significantly increases the value of cloud-averaged heating rates in all cases.

3.3 Summary

We have established, through a search of the literature about cloud microphysics and electromagnetic wave propagation, that a large-particle distribution of droplets is often superimposed on small-particle droplet distributions normally associated with nonprecipitating clouds. A method for computing the radiation attenuation parameters for a bimodal droplet distribution was developed. The attenuation parameters for a number of large-particle size distributions were tabulated and the bulk radiative characteristics of clouds composed of these distributions were presented and analyzed.

The large-particle droplet distributions result in significantly larger absorption values than did the small-particle distributions reported in Chapter 2. It is clear from these results that the limiting value of absorptance for water clouds may be greater than the 20% value previously reported.

Many combinations of small- and large-particle size distributions were used to produce bulk radiative characteristics of clouds. The variation of these characteristics was examined to determine the radiative effects of the bimodal distributions. While cloud liquid water content was primarily determined by the large drops, cloud radiation characteristics were somewhat insensitive to the actual size distribution of large drops, and more sensitive to the size distribution of small drops. These conclusions were valid only for situations in which large concentrations (10^2 cm^{-3}) of small drops were present. When the concentration of small drops was depleted to values of approximately 1 cm^{-3}, independent of size distribution, their effect upon the cloud radiation field was negligible in thick clouds.

For clouds of small optical depth ($\lesssim$ 30–40) and bimodal size distributions, the presence of small droplets increased the scattering field and increased drop absorption. However, for clouds of large optical depth ($\gtrsim$ 50), small droplets scattered more radiation from the cloud, thereby decreasing cloud absorption. As the small-droplet concentration in the bimodal distribution was decreased from 10^2 to 1 cm^{-3} the reflectance decreased nominally by about 20%.

CHAPTER 4

The Solar Radiative Properties of Ice Clouds

RONALD M. WELCH AND STEPHEN K. COX

Numerous investigators have alluded to the potential importance of upper tropospheric clouds on a great variety of meteorological problems. Manabe and Strickler (1964) studied the response of a thermal equilibrium model which included radiation and a convective adjustment to the presence of cirrus clouds. Atmospheric radiative variations were shown to depend upon cloud height and infrared emissivity. Manabe and Strickler (1964), Möller (1943), Cox (1969) and Reynolds *et al.* (1975) show that clouds may be warmed by radiative convergence. Reynolds *et al.* show that direct absorption of solar radiation by clouds is an important tropospheric heat source.

Gille and Krishnamurti (1972) found that when radiational cooling by infrared radiation is introduced into a tropical disturbance model, radiative-dynamical interactions enhance the conditional instability of such disturbances. Likewise, Reynolds *et al.* show that differential vertical heating by shortwave radiation in low clouds acts to destabilize the atmospheric layer so as to enhance cloud growth. In contrast, very strong heating rates observed in high cirrus clouds not only warm the upper troposphere and inhibit warming at lower levels, but also tend to stabilize the entire atmospheric air column.

Fleming and Cox (1974) demonstrated that the effective shortwave optical thickness of a cirrus cloud is the primary factor controlling the radiational energy budget at the earth's surface, while the cloud's broadband infrared emissivity is the primary factor influencing the heat budget of the atmosphere. Cox (1969) showed that cirrus clouds exert a surface heating tendency in the tropics, but have a net cooling effect on the surface in midlatitudes. This combination enhances the latitudinal radiative energy gradient.

While the radiative properties of cirrus clouds are not striking as compared to water clouds, their frequency of occurrence in time, large areal extent and location high in the troposphere make them important features in the vertical transport of energy. Appleman (1961) reported that a nearly invisible but persistent cirrus cloud layer exists near the tropopause over large areas of the earth. Fleming and Cox (1974) reported that tropical cirrus usually range in thickness from thin wisps to layers of 5 km or more, with bases ranging from 9 to 13 km.

Zdunkowski and Pryce (1974) point out that there is a lack of information in the literature concerning cloud composition and the physical characteristics of these clouds. Furthermore, even less is known concerning the scattering and absorbing characteristics of ice crystals. They point out that it is probably not feasible to determine the scattering characteristics of an optically thin cirrus from an examination of the radiative intensities emerging from cloud top.

4.1 Ice crystal characteristics

Relatively little is known about the microphysical structure of upper tropospheric ice clouds; this void of information is understandable, however, in light of the many problems associated with microphysical sampling in the upper troposphere. Weickmann (1947) found that the bullet rosette was the primary crystalline form in convective cirrus clouds, while eroded single bullets, columns and plates were found in cirrostratus clouds. Heymsfield (1975) reported that predominant ice crystal types were polycrystalline bullet rosettes, single bullets, banded columns and plates, in that order. Heymsfield (1977) found that ice crystal habits showed strong dependence upon air temperature and vertical velocity. Bullet rosettes, capped columns and plates were found between -17 and -27°C, dendrites and plates between -7 and -17°C, and aggregates of single crystals at temperatures warmer than -7°C. For cirrus uncinus clouds Heymsfield suggested that particles are smallest in the upshear region and increase in size with height. In the downshear region particles were found to increase in size with decreasing height, also the total concentration of particles was the greatest near the top of the upshear region. Furthermore, particle sizes and concentrations were found to be a function of temperature with sizes decreasing and concentrations increasing with decreasing temperature; in contrast the ice water content was a function of dynamics, not temperature.

In contrast to strong convective regions, Heymsfield reports that cirrostratus clouds exhibit lower particle concentrations and ice water contents. Mean and maximum crystal lengths were found to increase from near cloud top to cloud base. Heymsfield notes that numerous small ice crystals were observed at the cloud

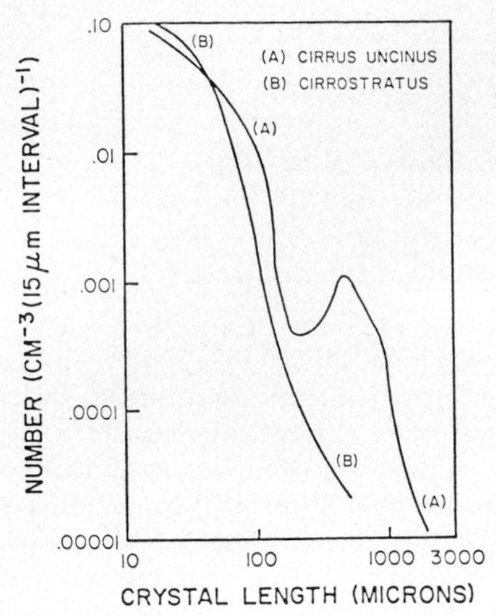

FIG. 4.1. Ice crystal size distributions from cirrus uncinus and cirrostratus measurements (from Heymsfield, 1975). Note the existence of the bimodal ice crystal size distribution present in the dynamically active case.

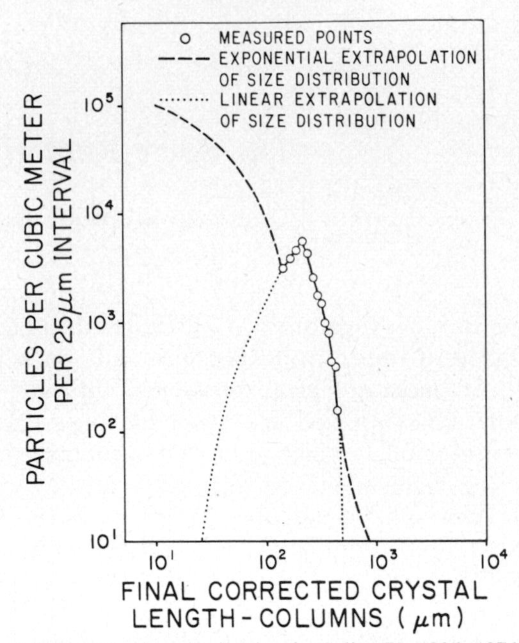

FIG. 4.2. Particle size distribution for Julian Day 226, 1974 at 10.4 km. Crystals are here assumed to be 100% columns. Measured points and two possible extrapolations are illustrated. (From Griffith *et al.*, 1980).

tops, indicative of new nucleation processes. However, due to limitations in the sampling probe, small-particle ice crystal concentrations were not observed. Heymsfield (1977) reported that ice water content and ice crystal concentration were found to be directly dependent on vertical velocity.

Fig. 4.1 shows ice crystal size distribution measurements from cirrus uncinus and cirrostratus clouds (from Heymsfield, 1975). Note the existence of the bimodal ice crystal size distribution present in the dynamically active case. As in Chapter 3, which considered the radiative properties of bimodal drop size distributions for water clouds, the present investigation considers the effects of bimodal ice crystal size distributions upon the radiative properties of cirrus clouds.

Fig. 4.2 shows the ice crystal size distribution measured by Griffith *et al.* (1980). It was observed that this size distribution measured by the one-dimensional Particle Measuring System, Inc. probe did not vary substantially from day to day or at different levels on the same day. The original measured size distribution is monomodal as shown. However, it was noted that a significant number of small particles may have been present below the lower sampling limit on the probe. A secondary, i.e. bimodal, size distribution of particles is theorized as shown.

Flux emissivities in cirrus clouds as measured by Griffith *et al.* are substantially greater than those predicted using the equivalent 50 μm sphere model by Hunt (1973) or by the ice column model of Liou (1974). However, both of these theoretical studies consider only a single particle size rather than a size distribution. It is well known (Deirmendjian, 1969) that this kind of approach may lead to serious errors in Mie scattering calculations. However, the primary conclusion reached by Griffith *et al.* is that differences between water clouds and ice clouds may not be as great as indicated by theoretical studies. The primary purpose of the present chapter is to explore the radiative characteristics of cirrus clouds in greater detail.

4.2 *Theoretical considerations*

Previous studies in Chapters 2 and 3 have considered interactions of the solar radiation field with water clouds. The present chapter may be considered to be an extension of this work in that the effects of monomodal and bimodal ice crystal size distributions upon the radiation field in cirrus clouds are examined. Water vapor transmission is approximated through the use of a sum of exponentials (Wiscombe and Evans, 1977; Welch *et al.*, 1976) based upon expansion coefficients taken from Liou and Sasamori (1975). Absorption by CO_2, ozone and trace gases is small and has been neglected in the present investigation.

Details of the method of calculation were presented in Chapters 2 and 3. To reiterate, the spherical harmonics method with the four-term expansion, developed by Zdunkowski and Korb (1974), has been converted into the delta-spherical harmonics technique applying the delta-M phase function expansion (Wiscombe, 1977). For large optical depths ($\tau > 35$), the adding method using the diamond initialization (Wiscombe, 1976) has been employed. Intercomparisons of radiative fluxes and heating rates between these two radiative transfer techniques show excellent agreement (Chapter 2). The vertical water vapor profile taken from GATE (Phase III) data has been used in most of the calculations. However, the tropical water vapor profile given by McClatchey *et al.* (1971) has also been used. In Chapter 2 it was shown for low-lying clouds that the choice of water vapor profile could lead to differences in cloud absorptance values and heating rates of up to 20%. However, for upper level clouds no significant differences are expected due to the small water vapor path length above the cloud.

4.2.1 Radiative properties of ice crystals

Calculations of the wavelength-dependent Mie efficiency factor $Q_\lambda(r)$ along with various crystal size distributions $n(r)$ combine to provide ice crystal absorption (β_a), scattering (β_s) and extinction (β_e) parameters [see Eq. (2.2)]. However, it is well known that ice crystals are not spherical in shape (Weickmann, 1947; Heymsfield and Knollenberg, 1972; Ryan *et al.*, 1972; McTaggart-Cowan *et al.* 1970; Braham and Spyers-Duran, 1967; Rosinski *et al.* 1970; Griffith *et al.*, 1980). Hunt (1973) has performed theoretical calculations using ice spheres of radius 50 μm. Liou (1972a,b; 1974) has extended the theoretical development to ice columns. Only in the theoretical work of Liou (1972a) has the effect of ice crystal size distributions upon the radiation field been examined.

Liou (1974) notes that differences in attenuation parameters between ice columns and equivalent volume ice spheres are small, but noticeable. He shows that the cloud optical depth in the infrared window region would be underestimated by a factor of 1.5 if spheres of 50 μm radius were assumed rather than cylinders of length 200 μm and radius 30 μm. If the radii of spheres are increased so that larger extinction cross sections are obtained to match the cylinders, then there is somewhat poorer agreement in the single-scattering albedos since larger spheres absorb more incident radiation.

Jacobowitz (1971) showed that the major difference between the scattering pattern of a hexagonal prism and a sphere, whose size was such as to scatter out of the incident beam an equal amount of energy, is the absence of the halo. For this case it was shown that the energy scattered in the angular region of the halo is comparable in magnitude to that of the forward peak. Jacobowitz (1970) reported that for ice crystals the scattering pattern due to diffraction can be approximated by the pattern of equivalent spheres while the reflection pattern is identical to that of the equivalent spheres. However, refraction was shown to be strongly dependent upon crystal shape.

While Liou (1972a,b; 1974) has assumed randomly oriented cylinders. Griffith *et al.* (1980) and Ono (1969) report that fluid dynamical considerations dictate that cylindrical crystals fall with the long axis horizontal. Furthermore, observations of lidar backscatter from high cirrus layers show strong angular dependence which indicates a coherent crystal orientation (Platt, 1977, 1978). The presence of optical phenomena, such as halos, confirm that there is some uniformity in crystal orientation, at least occasionally. Liou (1972b) has considered polydisperse ice crystals both in uniform orientation and in random orientation in a horizontal plane. For these cases it was once again found that cylinders scatter more light for scattering angles of $\sim$ 20–150° as compared to spheres, with decreased scattering in the forward and backward directions.

Measurements by Griffith *et al.* (1980) indicate that in cirrus clouds the large irregular particles of real cirrus clouds have significantly greater extinction efficiencies than either the spheres of Hunt's (1973) study or the regular columns of Liou's (1974) study. This may, in part, be due to the fact that realistic crystal size distributions have not been included in the theoretical calculations.

Measurements of ice crystals indicate that these particles are highly irregular in character (Weickmann, 1947; Heymsfield, 1975, 1977; Griffith *et al.*, 1980; and many others). Therefore, it may be the case that neither spheres, columns nor any other symetrically shaped object is a reliable model for the attenuation properties of ice crystals.

Scattering properties of spheres and irregular particles have been examined by Donn and Powell (1963), Pinnick *et al.* (1976), Napper and Ottewill (1963), Hodkinson (1963, 1966), Holland and Draper (1967), Holland and Gagne (1970), Zerull (1976) and Chýlek *et al.* (1976). Hodkinson (1966) shows that the extinction cross section may be approximated as twice the geometric cross section for large irregular particles or spheres. He shows that this approximation is valid for large size parameters

$$x > \frac{10}{|m - 1|},$$

where m is the index of refraction. Chýlek (1975) proved that the Mie efficiency for extinction approaches the value of 2 for large-particle size parameters. Hodkinson (1963), however, shows that with smaller-particle size parameters extinction coefficients are quite different between irregular particles and spheres.

Pinnick *et al.* (1976) find that for slightly nonspherical particles with size parameters (defined as the ratio of particle circumference to wavelength) greater than about 5, the intensity of light scattered in the forward peak was greater than that predicted by Mie theory, and less than at non-forward angles. For particles with size parameters < 5, results using Mie theory more closely matched the measurements. In addition, it was found that measured angular scattering patterns for randomly oriented particles were smoother than Mie theoretical results would predict.

Holland and Draper (1967) and Holland and Gagne (1970) report that the mass scattering coefficient for a polydisperse system of irregular but randomly oriented particles shows a remarkable similarity to the corresponding coefficient for spherical particles. However, they also show that backscattering is significantly different in these two cases.

Stephens *et al.* (1971) show that the far-field transient backscattering of electromagnetic radiation by liquid water spheres is principally the result of surface interactions. For ice spheres Ray and Stephens (1974) show that internal waves and diffracted fields are the principal contributors to scattering. It is shown that a combination of the dipole field with glory, axial, and stationary rays along with an internal surface wave of one cut describe the transient response for ice. Surface waves are excited by rays of grazing incidence which propagate along the surface of the particle. Due to symmetry considerations it is reasonable to assume that such surface waves are present in columns as well as in spheres. Van de Hulst (1957) suggested that surface waves can take a number of shortcuts through the sphere (or column), each time increasing the optical path length. Inada (1974) shows that such surface waves have an effective propagation constant closer to that of a dielectric sphere than to that of free space. A significant amount of the surface wave backscattering comes from surface waves which have made the maximum possible number of shortcuts through the sphere.

Chýlek (1976) has examined the resonances in partial waves obtained in Mie calculations and finds that these resonances are responsible for a ripple structure in the normalized extinction cross section. These results show that resonances in the Mie amplitudes are due to surface waves, and that absorption reduced these resonance peaks. Chýlek hypothesizes that the partial wave resonances are a function of photons traveling a long way along a surface of a sphere. Furthermore, he conjectures that such resonances cannot exist if the scattering particles have shapes drastically different from spheres or other highly symmetrical shapes. Therefore, he proposes that the first-order correction to nonsphericity should be a removal of partial-wave resonances.

Chýlek *et al.* (1976) set the Mie scattering functions to $a_n = 1/2$ or $b_n = 1/2$ in a resonance region for $n \geq 3$, where n is the partial-wave summation index, in order to remove these resonances. Scattering intensities calculated using this method give reasonable agreement with measured quantities. The concept of resonance breaks down for small-crystal size parameters since a surface wave can only exist on a spherical particle if the size of the sphere is larger than the wavelength of the incident radiation. It is emphasized that this approach should only be used for ensembles of randomly oriented and arbitrarily shaped nonspherical particles. It should also be noted that particle radii are to be determined from equivalent areas rather than from equivalent volumes.

Welch and Cox (1978) examined the nonspherical resonance approach as a function of size parameter and variations of the real and imaginary parts of the complex index of refraction. Nonspherical absorption efficiency was found to peak in the size parameter range $x = 5-10$, almost independent of the value of the imaginary component (n_i) of the refractive index. Differences in absorption efficiency between spherical and nonspherical particles increased (up to several orders of magnitude) for decreased values of n_i. Increased scattering (n_r) shifted the absorption peak to lower values of x and decreased the width of the absorption peak, but not the magnitude. These calculations were extended to show that there is a direct correspondence between decreased scattering intensity and increased absorption efficiency. For those values of size parameter in which absorption efficiency increased, there were decreases in both scattering efficiency and scattering intensity, particularly for large scattering angles. For values of size parameter in which there was no increased absorption efficiency, similarly there were no increases in scattering intensity.

Liou (1974) gives calculated values of extinction coefficients (β_e) and the single particle scattering albedo ($\tilde{\omega}_0$) at several wavelengths in the infrared window region for cylinders and spheres. The cylinders were assumed to be randomly oriented of length 200 μm and radius 30 μm. Table 4.1 gives comparable values for spheres using both the spherical and nonspherical approaches at a wavelength of 10 μm. Liou used comparisons with equivalent spheres of radius 50 μm. However, spheres of equivalent volume have radii of 51.3 μm and slightly larger attenuation parameters, although still considerably smaller than the values calculated for the ice crystals. For spheres of equivalent area (assuming horizontally oriented crystals), the equivalent sphere radius is 61.8 μm with extinction values very similar to those calculated by Liou. His results show increased scattering over that which would be obtained from Mie theory, while the Chýlek nonspherical theory predicts decreased scattering and increased absorption. The present investigation for cirrus clouds considers both spherical

TABLE 4.1. Extinction (β_e) coefficients (km^{-1}) and single-scattering albedos ($\tilde{\omega}_0$) for equivalent spheres at a wavelength of 10 μm. Liou's (1974) ice cylinders of length 200 μm and radius 30 μm are used for comparison. Spheres of equivalent volume have radii of 51.3 μm while spheres of equivalent area (horizontally oriented) have radii of 61.8 μm.

Type	Spherical		Nonspherical	
	β_e	$\tilde{\omega}_0$	β_e	$\tilde{\omega}_0$
Liou ice crystals			1.387	0.520
r = 50 μm	0.917	0.506	0.903	0.481
r = 51.3 μm	0.967	0.506	0.955	0.483
r = 61.8 μm	1.391	0.502	1.384	0.486

Mie calculations and those nonspherical corrections proposed by Chýlek *et al.* Considering the possible ice crystal shapes, size distributions and orientations, this approach appears to be a reasonable and economical compromise for the situation existing in cirrus clouds.

4.2.2 PHYSICAL PROPERTIES OF ICE CLOUDS

Heymsfield and Knollenberg (1972) reported that in dynamically active cirrus clouds, average values of ice crystal concentration for crystals longer than 150 μm were 0.1–0.25 cm^{-3}. The mean length of these crystals was 0.6–1.0 mm and the ice water content was 0.15–0.25 g m^{-3}. Heymsfield (1975) reported that crystal concentrations of 0.5 cm^{-3} existed in cirrus uncinus heads with concentrations of 0.025–0.05 cm^{-3} for particles longer than 100 μm. The mean length of crystals longer than 100 μm ranged between 0.2–0.5 mm. The ice water content for these clouds ranged between 0.01–0.16 g m^{-3}. Crystal sizes decreased and concentrations increased with decreasing temperatures.

Heymsfield (1977) found that ice water content generally increased from about 10^{-3} g m^{-3} at a temperature of −60°C to 0.5 g m^{-3} at 0°C. At temperatures >−15°C, ice crystal concentrations were 2–4 orders of magnitude higher than would be predicted from ice nuclei concentrations, in agreement with observations by Hobbs (1974). The mean and maximum crystal length was found to be only slightly dependent upon air temperature but directly dependent upon ice water content.

Ryan *et al.* (1972) measured ice crystal concentrations of 0.6–3.8 cm^{-3} for cirrus at 10 km and temperature of −43°C. McTaggart-Cowan *et al.* (1970) observed a maximum concentration of 0.53 cm^{-3} at −55°C and a concentration of 0.12 cm^{-3} at −29°C in the precipitation trail of a cirrus uncinus. Rosinski *et al.* (1970) found the ice water content to be 0.4–0.5 g m^{-3} at −33°C in cirrocumulus; 0.2 g m^{-3} at −33°C for cirrus; 0.005–0.1 g m^{-3} at −32°C to −37°C for thin cirrus; and 0.1 g m^{-3} for cirrus. Griffith *et al.* (1980) reported an ice water content of 0.1 g m^{-3} for a cirrus cloud between 200–300 mb and temperatures ranging between −30 and −60°C. On two different occasions Griffith *et al.* observed ice water content values of 0.02 and 0.025 g m^{-3} at cloud top and 0.12 and 0.015 g m^{-3} at cloud base; they reported mode crystal lengths varying from 200–250 μm with crystal sizes of up to 500 μm. Braham and Spyers-Duran (1967) reported ice crystal concentrations of 0.1–1.0 cm^{-3} several kilometers below cirrus clouds.

The foregoing review shows that microphysical properties are highly variable in cirrus clouds. Furthermore, only a few crystal size distribution measurements are available. Due to limitations in the instrumentation, no reliable measurements are available for small particles. In addition, the Particle Measuring System, Inc. one-dimensional probe is highly sensitive to particle shape (Knollenberg, 1975).

As mentioned in Chapter 3, the measurements of Sartor and Cannon indicate that ice particles may be prevalent in large quantities within regions of precipitation and strong convection. Their results are striking because they indicate that ice concentrations of 400 ℓ^{-1} and frozen water contents of 7 g m^{-3} may not be uncommon. Within such precipitation shafts graupel was prevalent. For such extreme conditions spherical Mie theory is quite sufficient, and nonspherical approximations do not need to be invoked. However, current knowledge of ice crystal, hail and snow size distributions is exceedingly limited. Therefore, even for spherical particles, assumed particle size distributions were used to derive most of the results presented in this chapter. In order to show the applicability of this approach, a large number of assumed and measured ice crystal size distributions are considered. Bulk radiative cloud characteristics (radiative fluxes, reflectance values, absorptance values and heating rates) will be shown to be somewhat insensitive to the choice of particle size distribution. This is an extremely useful result because taking into account the precise shape and orientation of each crystal in a cirrus cloud when making radiative transfer calculations would likely overburden even the largest computers. The fact that the bulk radiative characteristics of cirrus clouds are relatively insensitive to particle size distribution suggests that such complications may be avoided.

4.2.3 ICE CRYSTAL SIZE DISTRIBUTIONS

Measurements by Heymsfield (1975) show that large concentrations of small particles are likely to exist in cirrus clouds (Fig. 4.1). Results presented by Griffith *et al.* (1980), Varley (1978a, b), Varley and Brooks (1978) and Varley and Barnes (1979) also suggest the presence of small particles. Thus, we will consider a bimodal distribution of sizes. However, since the distribution of particle types is also unknown, a proper determination for the equivalent spherical size distribution is difficult (or impossible). As a result, several representative equivalent sphere sizes were chosen.

The C.5 and C.6 (Deirmendjian, 1975) size distributions have been used to represent a reasonable variation of small-crystal size distributions. Such distributions are naturally only representative of frozen cloud droplets, while real ice crystals take on a variety of shapes. Nevertheless, these size distributions span a reasonable range of particle sizes and provide an estimate, admittedly crude, of the effect of small particles on the radiative characteristics of cirrus clouds. As shown previously [Table 4.1, and Eqs. (4.2a) and (4.2b) to be discussed later], typical ice crystal length is approximately a factor of 3–5 times that of the equivalent sphere radius. At this point it should be reiterated that sphere radii determined from equivalent areas rather than equivalent volumes are used. The equivalent sphere radius will then depend upon crystal shape, i.e., width. However, such dimensions appear to be highly sensitive to particle length and to local equilibrium temperature. Therefore, actual size distributions may be highly variable. The results of Chapters 2 and 3 showed that for water drops, highly variable drop size distributions provide similar values for cloud bulk radiative characteristics, particularly for absorptance. Since the indices of refraction for water and ice are similar, equivalent behavior is expected for ice crystals.

While the previous discussion attempts to justify the use of arbitrarily chosen small-crystal size distributions, there is also considerable ignorance concerning the concentrations of such particles. The results for cirrus uncinus shown in Fig. 4.1 indicate that the small-particle concentrations are at least three orders of magnitude larger than those for the large particles. The small-particle concentrations will be scaled between 1 and 100 cm^{-3} to account for a reasonable variation of this parameter. The results in Table 4.1 suggest that small-crystal mode lengths of 10–20 μm may not be uncommon. Crystals of this length may correspond to size distribution functions such as the C.5 function with a mode radius of about 5 μm. It is interesting to note that even in cirrostratus clouds crystals in this size range are prevalent.

The Rain L (Deirmendjian, 1969), Rain 10 and Rain 50 (Deirmendjian, 1975) size distributions have been used for the large ice crystals. The mode radius of these distributions is 70, 330 and 600 μm, respectively. Assuming a factor of about 4 between crystal length and sphere radius, these distributions can also represent ice crystal size distributions which have crystal mode lengths of about 275, 1000 and 2000 μm, respectively. Therefore, the Rain L size distribution may be expected to give a reasonable approximation to the large-particle mode shown in Fig. 4.1. Likewise, the Rain L size distribution might be expected to reasonably approximate the size distribution shown in Fig. 4.2.

In addition, we shall let the Rain L, Rain 10 and Rain 50 size distributions represent crystal lengths rather than sphere radius. In such a case, the equivalent sphere radii for such distributions would be approximately 20, 80 and 150 μm, respectively.

In addition to these arbitrary ice crystal size distributions, the measurements shown in Fig. 4.2 have been fit to the modified gamma distribution function [Eq. (2.3)]. However, such a fit is somewhat arbitrary. Therefore, two different fits have been made, one with a mean crystal length of $L = 175$ μm, and the other with $L = 200$ μm.

If it can be established that the results for the $L = 175$ and $L = 200$, as well as the Rain L, distributions provide similar values of radiative characteristics, then one would be led to the conclusion that the actual choice of size distribution is not of primary importance.

In any case, it is the intention of the present investigation to analyze the impact of large- and small-particle sizes upon the radiation characteristics of cirrus clouds applying both spherical and nonspherical (resonance suppression) Mie calculations. As more complete radiation flux and microphysical measurements become available, more detailed computations can be performed. However, the present study will illuminate several important characteristic effects of ice crystals upon the ice cloud radiation field.

4.2.4 Complex indices of refraction for ice particles

Irvine and Pollack (1968) provide the complex indices of refraction for water and ice. Table 4.2a shows a comparison of ice and water indices of refraction

TABLE 4.2a. Indices of refraction for ice at selected wavelengths (from Irvine and Pollack, 1968). (Notation, e.g., 8–09 = $8*10^{-9}$).

λ	Water		Ice	
(μm)	n_r	n_i	n_r	n_i
0.35	1.349	8–09		
0.55	1.334	1.5–09		
0.75	1.329	1.49–07		
0.95	1.327	4.76–06	1.302	8.3–07
1.15	1.3235	7.32–06	1.299	2.93–06
1.30	1.321	11.17–06	1.296	1.24–05
1.50	1.318	2.065–04	1.294	5.58–04
1.80	1.312	1.146–04	1.292	1.13–04
2.0	1.304	1.082–03	1.291	1.61–03
2.5	1.246	1.651–03	1.235	7.95–04
2.8	1.232	9.361–02	1.152	0.0123
3.0	1.351	0.2586	1.130	0.2273
3.5	1.423	9.3–03	1.422	0.0163
4.0	1.349	4.81–03	1.327	0.0124
5.0	1.331	1.225–02	1.247	0.0133
6.0	1.313	0.010208	1.235	0.0617

at selected wavelengths. Although the values for ice are somewhat smaller than those of water over many wavelengths in the solar spectrum, in general these values of refractive index are so similar one would also expect similar values of attenuation parameters for equivalently sized water and ice particles.

The solar spectrum has been divided into a series of bands (Welch *et al.*, 1976). Within each region (Table 4.2b) the indices of refraction for ice, $m = n_r - in_i$, have been determined by weighting these values by the solar blackbody function at a temperature of $T = 5784°K$ (Strand, 1963), i.e.,

$$m = \frac{\int_{\Delta\lambda} m_\lambda B_\lambda (T = 5784 \text{ K}) d\lambda}{\int_{\Delta\lambda} B_\lambda (T = 5784 \text{ K}) d\lambda}. \quad (4.1)$$

A similar procedure has been carried out for water droplets by Zdunkowski *et al.* (1967). Since the value of n_i for ice is less than that for water, ice particles will absorb less radiation than corresponding water particles. Refractive indices for ice are not available at wavelengths < 0.95 μm (Irvine and Pollack, 1968; Schaaf and Williams, 1973; Bertie *et al.*, 1969). Therefore, values for water were used in these regions. The real parts of the complex refractive index for ice and water are assumed to be similar. However, the imaginary part for ice may be as much as a factor of 2 or 3 smaller than that for water. While it is realized that some error is unavoidable in this situation, the small values of the imaginary component of the refractive index for wavelengths < 0.95 μm make large errors in the bulk cloud radiative properties unlikely.

In order to determine the magnitude of such variations, n_i was divided by a factor of 3 and Mie calculations performed in the wavelength regions 0.75 and 0.55 μm. For both spherical and nonspherical particles, the extinction coefficient remained invariant to the choice of the value of n_i. Using the C.5 distribution for spherical particles, the absorption coefficient (β_a) has values (km^{-1}) of $9.33*10^{-4}$ and $1.17*10^{-4}$ at wavelengths of 0.75 and 0.55 μm, respectively. Decreasing the value of n_i by a factor of 3 leads to corresponding values of β_a of $3.11*10^{-4}$ and $3.90*10^{-5}$, respectively. Therefore, the value of β_a is nearly linear with the value of n_i in this wavelength region.

TABLE 4.2b. Averaged indices of refraction for ice using the solar blackbody function at $T = 5780$ K as a weighting function. (Notation, e.g., 1.2076–08 = $1.2076*10^{-8}$).

Spectral interval (μm)	n_r	n_i
0.3–0.7	1.3045	1.2076–08
0.7–0.8	1.3290	1.1462–07
0.8–1.0	1.3201	5.7748–07
1.0–1.2	1.3003	2.8447–06
1.2–1.6	1.2954	1.3973–04
1.6–2.2	1.2912	4.5083–04
2.2–3.0	1.1984	5.7743–02
3.0–3.57	1.4217	1.5277–01
3.57–8.69	1.2777	2.2009–02

The inherent uncertainties in size distributions, particle concentrations and particle shapes are undoubtedly far more important to the determination of cloud radiative characteristics than are small errors in the chosen values of refractive index. The assumption of water values for the complex index of refraction was considered more satisfactory than the choice of $n_i = 0$ as is assumed by many authors (i.e., Liou, 1974). While it may be tempting to ignore particle absorption in these small wavelength regions, absorption efficiency may become significant for large-particle size distributions.

Applying resonance suppression theory to simulate nonspherical particles, values of β_a were $1.65*10^{-2}$ and $1.02*10^{-2}$, respectively. Such values are one to two orders of magnitude larger than for spherical par-

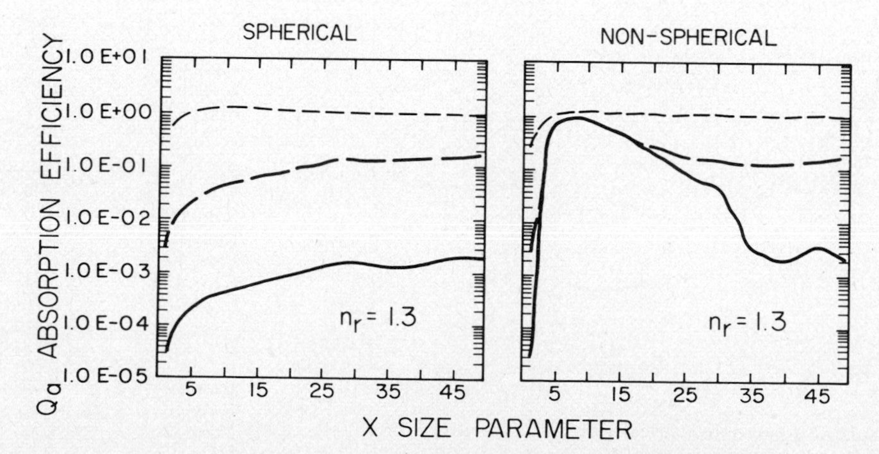

FIG. 4.3. Spherical and nonspherical Mie absorption efficiencies (Q_a) at small values of size parameter x for $n_r = 1.3$ and $n_i = 10^{-1}$ (----), 10^{-3} (– –), 10^{-5} (—).

ticles and decreasing the value of n_i by a factor of 3 had a negligible effect upon the value of β_a. Such behavior is totally unexpected and requires some comment.

Welch and Cox (1978) observed that resonance suppression theory leads to anomalously large values of Mie absorption efficiency (Q_a) for small values of the size parameter. Fig. 4.3 gives corresponding values of Q_a for $n_r = 1.3$ for both spherical and nonspherical particles. These results show that nonspherical absorption efficiency is nearly independent of the value of n_i for size parameters ranging from $x = 3$ to $x = 10$. For larger values of x the value of Q_a approaches that for spherical particles.

Fig. 4.4 shows spherical and nonspherical values of scattering intensity as a function of scattering angle for $n_r = 1.3$. These values have been integrated over the range in size parameter between $x = 8$ and $x = 12$. Scattering intensities are significantly smaller for nonspherical particles than for spherical ones. At larger values of x in which nonspherical Q_a approaches the spherical particle value, the nonspherical scattering intensity likewise approaches the spherical particle value.

The combination of Figs. 4.3 and 4.4, therefore, shows that the resonance suppression theory predicts a direct correlation between decreased particle scattering and increased particle absorption. Furthermore, such effects are limited to the smaller values of particle size parameter. Therefore, the effect of nonsphericity is strongly influenced by the particle size distribution. For particle sizes corresponding to size parameters < 30, radiative characteristics are particularly sensitive to nonspherical effects; however, for particles with $x \gtrsim 30$, no significant variations between spherical and nonspherical results are predicted using resonance suppression theory.

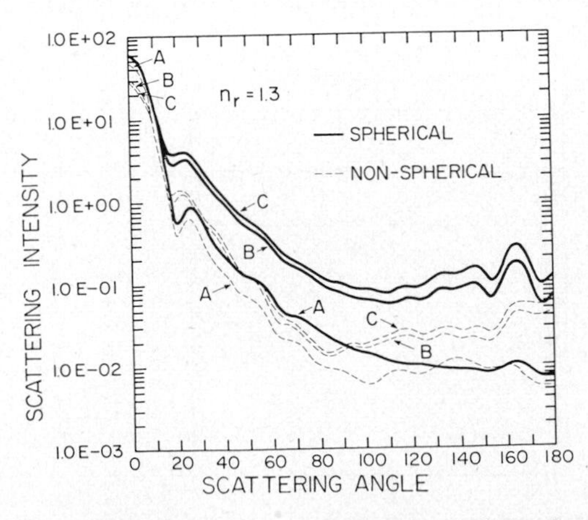

FIG. 4.4. Average spherical and nonspherical values of scattering intensity as a function of scattering angle for $n_r = 1.3$ and $n_i = 10^{-1}$ (A), 10^{-3} (B), 10^{-5} (C) for the range of size parameters between $x = 8$ and $x = 12$.

Resonance suppression theory is strictly valid only for arbitrarily shaped, randomly oriented particles, but most ice crystals have preferred orientation and shape. Therefore, resonance suppression theory must be applied with caution. Insofar as the present calculations are concerned, resonance suppression theory and spherical Mie theory can be considered as two extremes, with actual values falling somewhere in between the two results.

4.2.5 EXTINCTION, SCATTERING AND ABSORPTION PROPERTIES OF ICE CLOUDS

Table 4.3 lists ice water contents and ice crystal concentrations for the cases considered in the subsequent computations shown in Tables 4.4–4.7. Mie calculations assuming both spherical and nonspherical particles have been made, with equivalent sphere radii determined from equivalent particle cross-sectional area.

Extinction (β_e) and absorption (β_a) coefficients as a function of wavelength for each of the assumed spherical ice particle size distributions are given in Table 4.4. One notes that the increasing particle size leads to decreasing values of the single-scattering albedo [$\tilde{\omega}_0 = (\beta_e - \beta_a)/\beta_e$]. The range of applicability of these distributions has been discussed in Section 4.2.3. Comparison with Table 3.2 for water drops shows that water and ice have similar radiative characteristics.

In addition to the spherical drop size distributions, the $L_c = 175$ (Griffith *et al.*) distribution has also been used. For the purposes of the present calculation this

TABLE 4.3. Ice water contents and crystal concentrations for various particle size distributions of spheres, columns and bullets. (Notation, e.g., 1.30–02 = 1.30*10^{-2})

	Ice crystal size distribution	Ice water content (g m^{-3})	Ice crystal concentration (cm^{-3})
A: Spheres	C.5	0.296	10^2
	C.6	0.023	10^0
	Rain L	0.022	10^{-3}
	Rain 10	0.463	10^{-3}
	Rain 50	1.94	10^{-3}
	$L_c = 175$	0.053	8.8–04
B: Columns	C.5	1.30–02	10^2
	C.6	1.08–03	10^0
	Rain L	1.70–03	10^{-3}
	Rain 10	6.44–03	10^{-3}
	Rain 50	1.62–02	10^{-3}
	$L_c = 175$	1.51–03	8.8–04
	$L_c = 200$	2.76–03	1.32–03
C: Bullets	C.5	2.23–02	10^2
	C.6	9.60–04	10^0
	Rain L	1.17–03	10^{-3}
	Rain 10	4.86–03	10^{-3}
	Rain 50	1.37–02	10^{-3}
	$L_c = 175$	9.30–04	8.8–04
	$L_c = 200$	1.73–03	1.32–03

TABLE 4.4. Extinction (β_e) and absorption (β_a) coefficients (km^{-1}) along with the C_1 phase function expansion coefficients for various equivalent ice sphere size distributions as a function of wavelength (μm). (Notation, e.g., 1.17–04 = 1.17∗10^{-4})

Drop size distribution		Wavelength region (μm)								
		0.55	0.75	0.95	1.15	1.4	1.8	2.8	3.3	6.3
C.5	β_e	44.42	44.83	45.03	45.66	45.86	46.94	48.36	48.24	57.83
	β_a	1.17–04	9.33–04	3.37–03	1.36–02	5.40–01	1.35	23.18	23.96	23.41
	C_1	2.625	2.586	2.572	2.578	2.574	2.556	2.894	2.835	2.636
C.6	β_e	0.744	0.747	0.749	0.752	0.755	0.759	0.766	0.770	0.795
	β_a	8.60–06	5.97–05	2.43–04	9.79–04	3.63–02	8.12–02	0.372	0.358	0.362
	C_1	2.661	2.637	2.646	2.653	2.675	2.702	2.950	2.869	2.876
Rain L	β_e	0.365	0.366	0.366	0.366	0.366	0.367	0.368	0.369	0.371
	β_a	3.62–05	2.63–04	1.08–03	4.36–03	0.100	0.143	0.176	0.168	0.175
	C_1	2.602	2.533	2.625	2.610	2.730	2.817	2.959	2.887	2.928
Rain 10	β_e	1.295	1.295	1.295	1.295	1.295	1.295	1.30	1.30	1.31
	β_a	1.39–04	9.45–04	3.83–03	1.54–02	0.37	0.54	0.64	0.63	0.62
	C_1	2.527	2.517	2.594	2.569	2.691	2.779	2.939	2.843	2.937
Rain 50	β_e	3.53	3.53	3.53	3.54	3.54	3.54	3.55	3.55	3.56
	β_a	5.90–04	4.25–03	1.77–02	7.50–02	1.37	1.62	1.70	1.62	1.68
	C_1	2.652	2.597	2.642	2.618	2.753	2.820	2.930	2.826	2.926
L_c = 175	β_e	0.285	0.285	0.285	0.286	0.286	0.287	0.289	0.289	0.292
	β_a	1.48–05	1.00–04	4.09–04	1.67–03	5.26–02	9.55–02	0.138	0.133	0.138
	C_1	2.455	2.514	2.611	2.632	2.723	2.823	2.957	2.884	2.931

L_c = 175 distribution is assumed to describe particle radius rather than crystal length. The attenuation properties of this artificial construct are quite similar, however, to those obtained using the Rain L distribution. The purpose of these and following constructs is to gain familiarity with the effect of size distribution on attenuation parameters. In any case, there is no clear way to scale attenuation parameters with either particle concentration or ice water content.

As pointed out in Chapters 2 and 3, accurate radiative fluxes and cloud heating rates can be obtained using the Henyey-Greenstein phase function. The success of the delta-Eddington and the four-term delta-spherical harmonics methods stems from the fact that radiative fluxes are most strongly dependent upon the first few terms of the phase function expansion coefficients (Chapter 2; and Wiscombe, 1977). Therefore, the C_1 expansion coefficient is also given in Table 4.4.

Griffith *et al.* show that the assumed crystal shape alters the size distribution deduced from the Particle Measuring System, Inc. one-dimensional probe data. Thus when one assumes a bullet-rosette shape, a smaller modal crystal size results than when one assumes an ice column crystal configuration. Furthermore, the calculated ice water content depends on both particle shape and the size distribution function. For a given crystal size spectra, Griffith *et al.* observed that the calculated ice water content may be 0.703, 0.818 or 0.348 g m^{-3}, depending on whether the ice shapes are assumed to be spheres, columns or bullet rosettes, respectively. In the present work, both bullet-rosette and column crystal shapes will be assumed as we examine the radiative characteristics of ice clouds. We use the empirical relations [Eqs. (4.2a) and (4.2b)] from Heymsfield (1972) to relate crystal length (L) to crystal width (W) for the two assumed shapes:

$$\left.\begin{array}{ll} W = 0.25\,L^{0.7856}\ [\text{mm}], & L \leqslant 0.3\ \text{mm} \\ W = 0.185\,L^{0.532}\ [\text{mm}], & L \geqslant 0.3\ \text{mm} \end{array}\right\}\ \text{Bullets} \tag{4.2a}$$

$$\left.\begin{array}{ll} W = 0.5\,L\ [\text{mm}], & L \leqslant 0.2\ \text{mm} \\ W = 0.1973\,L^{0.414}\ [\text{mm}], & L \geqslant 0.2\ \text{mm} \end{array}\right\}\ \text{Columns} \tag{4.2b}$$

Size distributions with crystal model lengths of L_c = 175 μm and L_c = 200 μm from the data taken by Griffith *et al.* have also been used in the present calculations. The drop size distributions have been replaced so that they correspond to crystal length instead of sphere radius. The crystal size distributions are depicted as a function of crystal length instead of equivalent sphere radii, i.e.,

$$n(L) = a\,L^{\alpha} \exp\left[-\frac{\alpha}{\gamma}\left(\frac{L}{L_c}\right)^{\gamma}\right]. \tag{4.3}$$

The corresponding ice crystal concentration (cm^{-3}) is replaced by

$$N = \int_0^{\infty} n(L)\,dL. \tag{4.4}$$

Mie calculations for both spherical and nonspherical particles have been made, with equivalent sphere radii determined from equivalent particle areas,

$$r_{\text{eff}} = (LW/\pi)^{1/2}, \tag{4.5}$$

and the particle attenuation parameters are given by

TABLE 4.5. Extinction (β_e) and absorption (β_a) coefficients (km^{-1}) along with the C_1 phase function expansion coefficients for various equivalent area ice sphere size distributions as a function of wavelength. These values are for assumed ice columns using the Heymsfield (1972) relationship between length and width. Further discussion of this point is found in the text. (Notation, e.g., 8.70–06 = $8.70*10^{-6}$).

Drop size distri- butions		Wavelength region (μm) →								
		0.55	0.76	0.95	1.15	1.4	1.8	2.8	3.3	6.3
C.5	β_e	7.31	7.46	7.58	7.72	7.90	8.36	8.32	8.63	7.08
	β_a	8.70–06	5.84–05	2.49–04	9.61–04	3.83–02	9.17–02	3.07	4.20	1.03
	C_1	2.557	2.493	2.469	2.452	2.423	2.412	2.823	2.692	2.569
C.6	β_e	0.124	0.124	0.125	0.126	0.127	0.128	0.130	0.131	0.142
	β_a	6.07–07	4.25–6	1.93–05	7.13–05	2.73–03	6.42–03	6.43–02	6.31–02	5.04–02
	C_1	2.644	2.607	2.605	2.619	2.623	2.623	2.932	2.857	2.703
Rain L	β_e	3.87–02	3.88–02	3.89–02	3.91–02	3.92–02	3.93–02	3.95–02	3.97–02	4.03–02
	β_a	8.65–07	6.33–06	2.72–05	1.16–04	4.09–03	8.33–03	1.90–02	1.82–02	1.91–02
	C_1	2.551	2.586	2.667	2.672	2.741	2.785	2.954	2.870	2.924
Rain 10	β_e	0.119	0.119	0.120	0.120	0.121	0.121	0.122	0.122	0.124
	β_a	3.57–06	2.43–05	1.04–04	4.46–04	1.59–02	3.17–02	5.82–02	5.58–02	5.87–02
	C_1	2.627	2.549	2.653	2.643	2.758	2.829	2.955	2.869	2.931
Rain 50	β_e	0.248	0.248	0.248	0.249	0.249	0.251	0.252	0.252	0.255
	β_a	1.01–05	6.82–05	2.79–04	1.15–03	3.94–02	7.75–02	0.120	0.115	0.121
	C_1	2.516	2.527	2.622	2.605	2.750	2.840	2.961	2.882	2.932
L_c = 175	β_e	3.85–02	3.87–02	3.89–02	3.90–02	3.91–02	3.93–02	3.94–02	3.96–02	4.03–02
	β_a	7.51–07	5.83–06	2.41–05	9.92–05	3.50–03	7.31–03	1.89–02	1.82–02	1.92–02
	C_1	2.574	2.618	2.691	2.686	2.730	2.756	2.953	2.873	2.926
L_c = 200	β_e	6.74–02	6.77–02	6.80–02	6.82–02	6.83–02	6.86–02	6.89–02	6.92–02	7.04–02
	β_a	1.38–06	1.01–05	4.48–05	1.83–04	6.43–03	1.33–02	3.31–02	3.18–02	3.36–02
	C_1	2.631	2.592	2.693	2.686	2.743	2.766	2.953	2.873	2.927

TABLE 4.6. Extinction (β_e) and absorption (β_a) coefficients (km^{-1}) along with the C_1 phase function expansion coefficients for various equivalent ice sphere size distributions as a function of wavelength (μm). These values are for assumed ice bullets using the Heymsfield (1972) relationship between length and width. Further discussion of this point is found in the text. (Notation, e.g., 1.20–05 = $1.20*10^{-5}$).

Drop size distri- butions		Wavelength region (μm) →								
		0.55	0.76	0.95	1.15	1.4	1.8	2.8	3.3	6.3
C.5	β_e	9.83	10.01	10.17	10.33	10.50	10.89	11.58	11.45	10.85
	β_a	1.20–05	8.66–05	4.72–04	1.48–03	5.77–02	0.14	4.36	5.70	1.55
	C_1	2.570	2.511	2.489	2.477	2.443	2.403	2.834	2.722	2.601
C.6	β_e	0.118	0.119	0.120	0.120	0.121	0.123	0.125	0.126	0.136
	β_a	5.35–07	3.69–06	1.51–05	6.16–05	2.38–03	5.62–03	6.17–02	6.05–02	4.71–02
	C_1	2.644	2.607	2.602	2.616	2.619	2.620	2.931	2.856	2.689
Rain L	β_e	3.16–02	3.17–02	3.18–02	3.19–02	3.20–02	3.22–02	3.23–02	3.24–02	3.30–02
	β_a	6.73–07	4.90–06	2.11–05	2.88–05	3.17–03	6.48–03	1.55–02	1.49–02	1.56–02
	C_1	2.570	2.590	2.665	2.670	2.733	2.779	2.953	2.870	2.921
Rain 10	β_e	0.103	0.103	0.103	0.104	0.104	0.104	0.105	0.105	0.106
	β_a	2.86–06	2.01–05	8.63–05	3.72–04	1.33–02	2.63–02	5.01–02	4.81–02	5.06–02
	C_1	2.553	2.557	2.651	2.666	2.760	2.821	2.954	2.870	2.930
Rain 50	β_e	0.226	0.226	0.226	0.227	0.228	0.229	0.230	0.230	0.233
	β_a	8.84–06	6.05–05	2.48–04	1.04–03	3.57–02	6.98–02	0.110	0.105	0.110
	C_1	2.541	2.530	2.613	2.640	2.742	2.840	2.960	2.881	2.931
L_c = 175	β_e	3.00–02	3.02–02	3.03–02	3.03–02	3.04–02	3.05–02	3.07–02	3.08–02	3.14–02
	β_a	5.18–07	4.00–06	1.71–05	6.93–05	2.48–03	5.23–03	1.48–02	1.42–02	1.50–02
	C_1	2.575	2.623	2.688	2.680	2.717	2.750	2.953	2.872	2.923
L_c = 200	β_e	5.29–02	5.32–02	5.34–02	5.35–02	5.36–02	5.38–02	5.41–02	5.43–02	5.53–02
	β_a	9.52–07	7.40–06	3.19–05	1.30–04	4.61–03	4.41–02	2.60–02	2.50–02	2.64–02
	C_1	2.556	2.614	2.691	2.684	2.725	2.756	2.953	2.873	2.925

$$\beta_\lambda = \int_0^\infty n(L)\, \pi r_{\rm eff}^2\, Q_\lambda\, (r_{\rm eff})\, dL. \quad (4.6)$$

Table 4.5 gives attenuation parameters and the C_1 expansion coefficients assuming that the size distributions describe the length of ice columns. Such an assumption leads to significantly smaller values of equivalent sphere radii and, hence, significantly smaller values of attenuation coefficients. The choice of $L_c = 175$ or $L_c = 200$ for the Griffith *et al.* measurements (Fig. 4.2) leads to significant differences in the attenuation parameter. However, these differences are primarily a factor of particle concentration (Table 4.4).

Table 4.6 shows similar calculations assuming that the size distributions describe the length of ice bullets. Tables 4.4 and 4.5 show that the bullet-rosette crystals produce significantly larger values of attenuation coefficient than do ice columns. Furthermore, these differences are magnified for larger-particle sizes and are as large as a factor of 2.

Table 4.7 gives attenuation coefficients calculated using resonance suppression theory. Only the small-particle C.5 and C.6 distributions have been assumed in these calculations. Since the resonance suppression effects are limited to small-size parameters the larger particle size distributions are relatively unaffected.

Comparison of Tables 4.4 and 4.7 suggests that extinction coefficients (β_e) are relatively invariant to nonspherical resonance suppression and the effect upon the absorption coefficient (β_a) is highly dependent upon wavelength. At short wavelengths β_a may be two orders of magnitude greater than the spherical case values for the C.5 and C.6 distributions. However, at long wavelengths the effect is negligible because the absorption efficiency has already reached its maximum value.

4.3 Results

As discussed in the previous section, attenuation parameters vary with ice crystal size distribution as well as particle shape. The present section will be concerned with the radiative characteristics of ice clouds which result from these attenuation coefficients. Both monomodal and bimodal ice crystal size distributions will be considered. Since ice crystal concentrations and ice water contents are highly variable, the various attenuation parameters will be scaled to consider a wide range of possibilities. Also since ice crystals may coexist with water droplets, situations with both ice crystal concentrations and liquid cloud droplets will be simulated.

4.3.1 Spherical and Nonspherical Monomodal Small-Particle Distributions

Table 4.8 shows calculations of bulk cloud radiative characteristics for various monomodal, small-particle, ice crystal size distributions as a function of wavelength region. Cloud thickness is 3 km with base height of 9 km and solar zenith angle of $\theta = 0°$. Both spherical and nonspherical (resonance suppression theory) particles, using the C.6 spherical particle distribution, provide equivalent results. Resonance suppression theory is only applicable for particles with small-particle size parameters. The results of the C.6 (spheres) calculation indicate that applying the nonspherical approach to larger particle sizes will show negligible differences. These results do not categorically show that larger nonspherical particles have radiative properties similar to those of equivalent spheres; rather these results merely show that resonance suppression theory does not predict significantly different radiative characteristics (than equivalent spheres) for large particles.

TABLE 4.7. Extinction (β_e) and absorption (β_a) coefficients (km^{-1}) along with the C_1 phase function expansion coefficients for various equivalent ice sphere size distributions as a function of wavelength (μm). These values are given for nonspherical particles, applying the Chylek resonance suppression theory. (Notation, e.g., 1.02–02 = 1.02*10^{-2}).

Drop size distributions		Wavelength region (μm) 0.55	0.76	0.95	1.15	1.4	1.8	2.8	3.3	6.3
C.5	β_e	44.41	44.82	45.01	45.59	45.66	46.34	46.09	47.23	46.05
	β_a	1.02–02	1.65–02	6.89–02	0.258	1.11	2.99	24.38	24.88	22.66
	C_1	2.608	2.574	2.566	2.596	2.618	2.652	2.884	2.824	2.637
C.6	β_e	0.735	0.740	0.744	0.748	0.753	0.759	0.765	0.768	0.787
	β_a	1.11–04	4.96–04	2.12–04	8.99–04	3.49–02	8.11–02	0.373	0.360	0.373
	C_1	2.646	2.606	2.614	2.634	2.648	2.666	2.944	2.864	2.847
C.5 Bullets	β_e	9.82	9.97	10.05	10.01	9.85	9.36	9.87	10.05	9.31
	β_a	4.74–02	0.14	0.36	0.81	1.44	2.51	5.11	6.01	3.26
	C_1	2.587	2.562	2.598	2.653	2.668	2.651	2.768	2.694	2.363
C.6 Bullets	β_e	0.118	0.119	0.120	0.120	0.121	0.122	0.123	0.123	0.125
	β_a	4.08–06	1.52–05	4.42–05	1.02–04	2.63–03	6.38–03	6.31–02	6.25–02	5.94–02
	C_1	2.627	2.583	2.577	2.603	2.613	2.632	2.917	2.845	2.724

TABLE 4.8. Percent cloud reflectance *(R)*, transmittance *(T)* and absorptance *(A)* as a function of wavelength region for several monomodal small-particle ice crystal distributions. Cloud thickness is 3 km, cloud base height 9 km and solar zenith angle $\theta = 0°$.

Drop size distribution		Wavelength region (μm) 0.55	0.76	0.95	1.15	1.4	1.8	2.8	3.3	6.3	Total
C.6 (spherical & nonspherical)	*R*	7.4	8.0	7.3	7.0	3.9	2.5	0.1	0.3	0.2	6.3
	T	92.6	92.0	88.4	87.7	63.8	53.6	16.5	28.4	15.4	82.5
	A	0.0	0.03	4.3	5.3	32.3	43.9	83.4	71.3	84.4	11.2
C.6 (bullets and columns)	*R*	1.1	1.2	1.1	1.1	0.8	0.8	0.1	0.1	0.2	1.0
	T	98.9	98.8	95.2	94.6	77.0	73.2	42.5	72.0	40.7	91.6
	A	0.0	0.0	3.7	4.3	22.2	26.1	57.4	27.9	59.1	7.4
C.5 (spherical)	*R*	90.6	91.1	87.3	82.8	39.7	25.9	0.2	0.3	2.5	75.3
	T	9.3	8.3	6.3	4.2	0.0	0.0	0.0	0.0	0.0	6.1
	A	0.1	0.6	6.4	13.0	60.3	74.1	99.8	99.7	97.5	18.6
C.5 (nonspherical)	*R*	88.4	84.0	71.5	53.9	26.5	11.8	0.2	0.3	0.7	65.9
	T	7.6	4.5	1.1	0.1	0.0	0.0	0.0	0.0	0.0	3.8
	A	4.0	11.5	27.4	46.0	73.5	88.2	99.8	99.7	99.3	30.3
C.5 (bullets)	*R*	70.3	73.4	71.0	70.4	48.0	38.7	0.4	0.6	4.1	63.2
	T	29.7	26.5	23.1	21.3	7.9	3.2	0.1	0.0	0.1	22.0
	A	0.01	0.06	5.9	8.3	44.1	58.1	99.5	99.4	95.8	14.8
C.5 (columns)	*R*	64.1	67.8	65.8	65.7	46.9	38.9	0.5	0.7	4.2	58.7
	T	35.9	32.2	28.3	26.2	12.8	7.3	0.0	0.0	0.1	27.3
	A	0.006	0.04	5.9	8.1	40.3	53.8	99.5	99.3	95.7	14.0
C.5 bullets (nonspherical)	*R*	54.7	40.9	22.5	9.8	4.5	2.1	0.4	0.5	2.3	32.4
	T	19.6	9.0	2.5	0.5	0.1	0.0	0.0	0.0	0.0	9.7
	A	25.7	50.1	75.0	89.7	95.4	97.9	99.6	99.5	97.7	57.9
Nimbostratus top (nonspherical)	*R*	92.6	89.9	81.7	70.6	30.7	16.1	0.2	0.3	0.6	72.2
	T	3.0	1.4	0.3	0.1	0.0	0.0	0.0	0.0	0.0	1.5
	A	4.4	8.7	18.0	29.3	69.3	83.9	99.8	99.7	99.4	26.3
Nimbostratus top (bullets)	*R*	82.3	84.1	80.8	78.7	45.0	32.8	0.3	0.4	3.5	70.8
	T	17.7	15.7	13.3	11.9	1.0	0.1	0.0	0.0	0.0	12.4
	A	0.03	0.2	5.9	9.4	54.0	67.1	99.7	99.6	96.5	16.8
Nimbostratus top (columns)	*R*	79.2	81.3	78.3	76.4	45.4	33.8	0.3	0.4	3.6	68.8
	T	20.8	18.5	15.9	14.3	2.1	0.4	0.0	0.0	0.0	14.7
	A	0.02	0.2	5.8	9.3	52.5	65.8	99.7	99.6	96.4	16.5

Results for the C.6 distribution give an absorptance value of 11.2%, with a cloud-averaged heating rate of 0.49°C h^{-1}. Measurements by Reynolds *et al.* for cirrus clouds give absorptance values of 13–14%, in rough agreement. However, these measurements also provide reflectance values of 47–59%, values comparable to water clouds. The C.6 distribution provides a value of only 6.3%. As shown in Chapter 3, reflectance is determined more from particle concentration than from either liquid (or ice) water content or particle size. Therefore, much larger concentrations of particles may be required than provided by the C.6 distribution. Results obtained by varying particle concentrations will be discussed later.

As discussed in Section 4.2.5 the C.6 distribution of spherical particles may represent ice crystals with lengths about 3–4 times the sphere radius, or particles with lengths of about 60–80 μm. The C.6-bullets distribution represents particles with a mode length of 20 μm, and with width-to-length ratios given by Eq. (4.2). For these smaller particles cloud absorptance is reduced to 7.4% and reflectance to 1.0%. The value of absorptance for bullets is reduced from that of spheres in all wavelength regions because of the smaller particle size. The decrease in the value of reflectance appears to be roughly correlated with the value of scattering coefficient. Replacing columns for bullets yields equivalent results.

The next monomodal size distribution in Table 4.8 is the C.5 spherical. This distribution represents ice crystals of mode length 15–20 μm; due primarily to the large-particle concentration (10^2 cm^{-3}), cloud reflectance has increased to 75%, while absorptance has increased to 18.6%. Cloud-averaged heating rate has increased to 0.81°C h^{-1}. These values are larger than those measured by Reynolds *et al.* suggesting that Reynolds' observations were made for clouds with smaller particle concentrations.

The next distribution uses the same C.5 size spectra, but applies resonance suppression theory for arbitrarily shaped, randomly oriented particles. In this case there is a very large increase in cloud absorptance (to 30%) with a decrease in cloud reflectance (to 66%). Note that the percent increase in absorptance is largest at the smallest wavelengths. At wavelengths near 1 μm the increase in absorptance applying nonspherical theory is about a factor of 4; for wavelengths < 0.8 μm much larger increases occur. Of course, most ice crystals show preferred symmetry and preferred orientation, so that the resonance suppression theory cannot be rigorously applied. Nevertheless, these results provide an extreme case. Real particles presumably have values somewhere in between values given by the spherical and nonspherical limits.

The next size distribution applies the C.5 distribution function to the length of ice crystal bullets. This distribution provides smaller particle sizes than does the C.5 spherical case along with smaller values of cloud reflectance (63%) and absorptance (14.8%). Assuming that the C.5 distribution describes ice columns instead of ice bullets, provides slightly smaller values of cloud reflectance (59%) and absorptance (14%), resulting from the somewhat smaller size of crystal widths. Cloud-averaged heating rates for these cases are about 0.62°C h^{-1}.

Resonance suppression theory when applied to the C.5-bullet distribution predicts a cloud absorptance of 58%, a value which appears to be totally unreasonable. The very large value of absorptance in this case arises from the fact that this particle distribution has values of size parameter in which absorption is maximum (Fig. 4.2). While such absorption values appear at first to be unrealistically large these results do suggest that anomalous values of absorption may indeed be real for some small crystal sizes.

It was shown in Chapters 2 and 3 that bulk radiative characteristics of a cloud depend on the choice of small-particle size distribution. Therefore, the size distribution representing spherical particles found at the top of nimbostratus (NS) clouds has also been used for comparison purposes. As seen in Table 4.8, this NS top distribution for nonspherical particles gives comparable, although somewhat smaller, values of cloud reflectance and larger values of cloud absorptance than does the C.5 distribution. This comes about because this size distribution has fewer particles with values of size parameter which lie within the anomalous absorption region. However, the larger particle sizes of the NS top distribution provide both larger reflectance and absorptance values than does the C.5 distribution for bullets and columns.

Table 4.8 shows that for the same particle concentration (10^2 cm^{-3}) cloud reflectance values for the C.5 spherical, bullet and column cases vary between about 60–75% and that absorptance values may vary between 14–18%. Application of resonance suppression theory to these small-particle sizes leads to much larger values of absorptance. In any case, these results are quite similar to those obtained from comparable size distributions of water drops. A water cloud with the C.5 size distribution provides a cloud reflectance value of 76% and an absorptance value of 18.4%. These results support the measurements of Griffith *et al.* (1980) who found that ice clouds appear to have radiative characteristics similar to those of water clouds.

4.3.2 Monomodal large-particle size distributions

Table 4.9 gives the bulk cloud radiative characteristics for various monomodal large-particle size distributions in a cloud 3 km thick with solar zenith angle of 0°. Most of these distributions have particle concentrations of ~ 10^{-3} cm^{-3}; calculations are also given in which these concentrations have been increased by one and two orders of magnitude.

The first size distribution given is that for spherical particles represented by the Rain L size spectrum. Radiative characteristics of cylindrical particles with lengths ranging between 250–300 μm should be similar to those represented by this size distribution. A cloud reflectance value of only 3.2% results from the size spectrum; cloud absorptance is about 12.5% with a cloud averaged heating rate of 0.54°C h^{-1}. Increasing the particle concentration by an order of magnitude increases cloud reflectance to 32.8% and cloud absorptance to 26.5%, with a heating rate of 1.15°C h^{-1}. This absorptance value is larger than that reported by Reynolds *et al.*, while the reflectance value is approximately the same as the measured value. Increasing the ice crystal number density another order of magnitude increases cloud reflectance to 58.7% and cloud absorptance to 36.5%. The cloud reflectance value increases nearly linearly with both particle number density and optical depth in the range $N = 10^{-3}$ cm^{-3} to $N = 10^{-2}$ cm^{-3}. However, the rate of increase in the reflectance value for larger particle number densities/optical depth is smaller as the cloud optical depth increases. This behavior is a result of the fact that multiple-scattering effects tend to become "saturated", or nearly isotropic, as the cloud optical depth increases. Cloud absorptance increases approximately logarithmically with ice particle number density.

Selecting a larger particle size distribution such as the Rain 10 size spectrum provides larger values of both cloud reflectance and absorptance. Ice crystals with lengths of about 0.8–1.2 mm may be represented by this distribution. Cloud reflectance has increased to 14.1% with absorptance values of 21.2%; cloud-averaged heating rate with this distribution is 0.92°C h^{-1}. Increasing the particle concentration by an order of magnitude leads to a cloud reflectance value of 53.2% and a cloud absorptance value of 32.8%. While cloud reflectance is similar to that reported by Reynolds *et al.*, the cloud absorptance we compute is much larger.

TABLE 4.9. Percent cloud reflectance (R), transmittance (T), absorptance (A) and cloud-averaged heating rates [$\partial\theta/\partial t$ (°C h^{-1})] for several monomodal large-particle ice crystal size distributions. Calculations for particle concentrations increased by an order of magnitude are also given. Cloud thickness is 3 km, cloud base height 9 km and solar zenith angle θ = 0°.

Drop size distribution		Monomodal	$\frac{\partial\theta}{\partial t}$	Monomodal*10	$\frac{\partial\theta}{\partial t}$	Monomodal*100	$\frac{\partial\theta}{\partial t}$
Rain L (spheres)	R	3.2		32.8		58.7	
	T	84.3	0.54	40.7	1.15	4.8	1.59
	A	12.5		26.5		36.5	
Rain 10 (spheres)	R	14.1		53.2		61.3	
	T	64.7	0.92	14.0	1.43	0.3	1.67
	A	21.2		32.8		38.4	
Rain 50 (spheres)	R	26.4		54.6		56.7	
	T	44.6	1.26	5.2	1.75	0.0	1.88
	A	29.0		40.2		43.3	
Rain L (columns)	R	0.3		3.6		33.2	
	T	92.5	0.31	86.1	0.45	43.6	1.01
	A	7.2		10.3		23.2	
Rain 10 (columns)	R	0.9		10.8		52.4	
	T	91.0	0.35	72.5	0.73	20.1	1.19
	A	8.1		16.7		27.5	
Rain 50 (columns)	R	2.4		25.6		61.7	
	T	87.8	0.43	51.9	0.98	7.8	1.33
	A	9.8		22.5		30.5	
L_c = 175 (bullets)	R	0.3		2.6		26.9	
	T	92.7	0.31	88.1	0.40	52.3	0.90
	A	7.0		9.3		20.8	
L_c = 200 (bullets)	R	0.4		4.9		39.4	
	T	91.8	0.34	84.0	0.48	36.7	1.04
	A	7.8		11.1		23.9	
L_c = 175 (columns)	R	0.3		3.3		32.0	
	T	92.6	0.31	86.6	0.44	45.4	0.98
	A	7.1		10.1		22.6	
L_c = 200 (columns)	R	0.5		5.6		49.2	
	T	92.1	0.32	82.1	0.53	26.4	1.06
	A	7.4		12.3		24.4	

The Rain 50 size distribution represents very large ice crystals, with crystal lengths of about 1.75–2.5 mm. Ice water content for this distribution is about 2 g m^{-3} whereas Heymsfield reports ice water contents normally below 0.5 g m^{-3}. Nevertheless, in highly convective clouds Sartor and Cannon reported ice water contents of 7 g m^{-3}, a value similar to that given by the Rain 10 distribution with a concentration of 10^{-2} cm. In any case this large-particle Rain 50 distribution provides a cloud reflectance value of 26.4% and an absorptance value of 29%. Increasing the particle concentration by an order of magnitude leads to a reflectance value of 54.6% and an absorptance value of 40.2%. These values are of the magnitude measured in water clouds. However, ice water contents of this magnitude are not realistic, and these results are presented merely for comparison.

Allowing the spherical drop size distribution to represent ice crystal lengths instead of sphere radii, provides smaller particle sizes and smaller values of the bulk radiative cloud characteristics. Allowing the Rain L, Rain 10 and Rain 50 distributions to represent crystal lengths of columns, and using Heymsfield's length-to-width relationships [Eqs. (4.2a) and (4.2b)] lead to values of cloud reflectance of 0.3, 0.9 and 2.4%, respectively. These size distributions represent crystals with mode lengths of 70, 330 and 600 μm, respectively, with absorptance values of 7.2, 8.1 and 9.8%. Cloud optical depth and reflectance values increase rapidly with increasing particle size. The values of cloud reflectance increase much more slowly with increasing cloud optical depth. Increasing the ice particle number density by an order of magnitude leads to a similar increase in the value of cloud reflectance, to values of 3.6, 10.8 and 25.6%, respectively. Corresponding values of cloud absorptance increase to 10.3, 16.7 and 22.5%, respectively. The values of cloud absorptance measured by Reynolds *et al.* are similar to those given by the Rain L and Rain 10 distributions; however, the measured values of reflectance are considerably larger than those provided by these distributions. Increasing the particle number densities by another order of magnitude provides reasonable values of reflectance, but absorptance values which are much too large.

Measurements by Griffith *et al.* (1980) are represented by the L_c = 175 μm and L_c = 200 μm mode length distributions. Differences between reflectance and absorptance using these two distributions are slight. However, increasing the particle concentration by an order of magnitude increases the differences in radiative characteristics between these distributions. Both cloud reflectance and absorptance are larger for the L_c = 200 μm case than for the L_c = 175 μm case. Note that these values are quite similar to those obtained using the spherical Rain L distribution which represents crystal mode lengths of about 250–300 μm. These results show that (the nonspherical and spherical) distributions can yield essentially equivalent values of bulk radiative characteristics if the equivalent spherical radius is properly scaled. It appears that a size distribution which represents crystal length or equivalent sphere radii (from equivalent areas) may be a reasonable approximation for ice crystal length distributions. However, knowledge of length and width ratios and crystal type appears to be required as well. Calculations using the L_c = 175 μm and L_c = 200 μm size distributions for columns provide similar values to those given by bullets. Increasing the particle concentration by an order of magnitude leads to radiative characteristics which are still relatively similar to each other and similar to the Rain L distribution. Values of reflectance vary from about 2.6 to about 5.6% with the values of absorptance ranging from about 9 to 11.1%. Values of absorptance in this range are similar to those measured by Reynolds *et al.*; however, the corresponding values of cloud reflectance are much lower than their measured values. Increasing particle number density/optical depth/ice water content by another order of magnitude once again produces reasonable agreement with measured values of cloud reflectance, but poor agreement with measured values of cloud absorptance.

The above results for both large particle and small particle monomodal size distributions have been calculated for a single cloud thickness. The effect of cloud thickness upon radiative characteristics of cirrus clouds is considered in Table 4.10 which shows cloud reflectance and absorptance values as a function of cloud thickness for several monomodal ice crystal size distributions. Cloud base height is 3 km. Cloud reflectance is nearly independent of both cloud thickness and ice crystal size distribution, except for the large-particle spherical distributions. Only for the very large Rain 50 sphere distributions do larger values of cloud reflectance result. Values of cloud absorptance are also relatively independent of particle size distribution. For a 500 m thick cloud, absorptance values typically vary from 2.4 to 3.6% (excluding Rain 50 spheres); for a cloud with thickness of 1000 m, values range from 4.1 to 5.2%; while for a cloud with thickness of 2000 m, values range from 5.9 to 8.0%. Assum-

TABLE 4.10. Percent cloud reflectance (R) and absorptance (A) for three cloud thicknesses as a function of monomodal size distribution. Cloud base height is 3 km and solar zenith angle θ = 0°.

Drop size distribution		Monomodal: Cloud thickness 500 m	Monomodal: 1000 m	Monomodal: 2000 m	Monomodal*10: Cloud thickness 500 m	Monomodal*10: 1000 m	Monomodal*10: 2000 m
Rain L (spheres)	R	0.5	1.0	2.1	5.6	11.7	22.4
	A	3.6	6.3	9.9	12.4	19.0	24.7
Rain 50 (spheres)	R	4.7	9.8	19.2	36.4	46.6	52.6
	A	14.0	20.8	26.4	30.8	35.5	38.9
L_c = 175 (bullets)	R	0.1	0.1	0.2	0.4	0.8	1.7
	A	2.4	4.1	5.9	2.9	5.0	7.5
L_c = 200 (bullets)	R	0.1	0.2	0.3	0.8	1.5	3.2
	A	2.6	4.4	6.4	3.3	5.7	8.8
L_c = 175 (columns)	R	0.1	0.1	0.2	0.5	1.1	2.2
	A	2.4	4.1	6.0	3.1	5.3	8.1
L_c = 200 (columns)	R	0.1	0.2	0.3	0.9	1.7	3.6
	A	2.5	4.2	6.2	3.6	6.3	9.7
Rain L (columns)	R	0.1	0.1	0.2	0.6	1.1	2.3
	A	2.5	4.1	6.0	3.1	5.4	8.3
Rain 10 (columns)	R	0.2	0.3	0.6	1.6	3.3	7.0
	A	2.7	4.5	6.7	5.0	8.6	13.4
Rain 50 (columns)	R	0.4	0.8	1.6	4.3	9.1	18.1
	A	3.0	5.2	7.9	8.0	13.3	19.3

ing spherical ice particles with the Rain 50 distribution results in very large values of cloud absorptance, comparable to those obtained assuming water drops.

Increasing the ice crystal number density by an order of magnitude once again increases the values of cloud reflectance by about an order of magnitude for most size distributions for all of these cloud thicknesses. It is interesting to note that the values of cloud absorptance are generally larger than those of cloud reflectance for the large-particle sizes. This relation tends to persist even for the optically thicker clouds. Smaller particles tend to produce values of cloud reflectance which are much larger than the corresponding values of cloud absorptance, due in part to the much larger particle concentrations and much larger values of single scattering albedo. Nevertheless, as the small-particle concentrations are decreased, the value of cloud reflectance decreases much more rapidly than does the value of cloud absorptance. Decreasing the small-particle number densities by three to four orders of magnitude leads to a situation in which the value of cloud reflectance is once again smaller than the value of cloud absorptance.

The value of reflectance is directly related to multiple scattering and to the optical depth of the scattering field. Interestingly enough, particle absorption shows a much weaker dependence on the existent multiple scattering. This is particularly evident in high cirrus in which water vapor absorption is essentially negligible.

In any case, regardless of size distribution, complete agreement with the Reynolds *et al.* values of cloud absorptance and reflectance cannot be obtained. For most size distributions, the values of cloud reflectance are significantly smaller than those which they measured. Increasing particle concentrations leads to reasonable values of cloud reflectance, but also leads to much larger values of absorptance than those measured by Reynolds *et al.* The conclusion is that neither monomodal small-particle nor monomodal large-particle size distributions can be used to adequately model the bulk radiative characteristics of clouds reported by Reynolds *et al.* We shall, therefore, examine the radiative characteristics associated with bimodal distributions of both large and small particles.

4.3.3. Bimodal ice crystal size distributions

Table 4.11 shows values of cloud reflectance *(R)*, cloud absorptance *(A)* and cloud-averaged heating rates ($\partial\theta/\partial t$) for various bimodal ice crystal size distributions. The large-particle size distributions listed in the left-hand column are combined with the spherical C.6 small-particle distribution to form various bimodal size distribution combinations. Calculations for cloud thicknesses of 500, 1000 and 3000 m are given for a cloud base height of 9 km and solar zenith angle of $\theta = 0°$. A second set of calculations in which the large-particle mode is increased in concentration by an order of magnitude is presented also.

The C.6 + Rain L (spheres) size distribution for a cloud 500 m thick results in a cloud reflectance value of 1.4%, cloud absorptance 4.6% and cloud-averaged heating rate of 1.03°C h^{-1}. Doubling the cloud thickness to 1 km also leads to a doubling of the value of reflectance (2.9%), and a significantly larger value of absorption (8.0%). The bulk cloud-heating rate is, however, smaller (0.93°C h^{-1}) than that found for the thinner cloud. Tripling the cloud thickness to 3 km also triples the value of reflectance (to 9.4%) and doubles the value of absorptance (15.9%), while decreasing further the bulk cloud-heating rate (0.69°C h^{-1}). Similar values of bulk cloud radiative characteristics are obtained when the Rain L (columns), Rain 50 (columns and bullets), L_c = 175 (bullets and columns) or L_c = 200 (bullets and columns) distributions are substituted for the Rain L (spherical) distribution. Bulk cloud radiative characteristics are therefore relatively insensitive to the large-particle mode. However, for progressively larger particle sizes, the values of cloud reflectance and absorptance do increase. The Rain 50 (spherical) and Rain 10 (spherical) bimodal size distributions have much different radiative characteristics. The effective radii of bullets and columns as determined from Eq. (4.5) are a factor of about 3–7 times smaller than those given by the spherical distributions. These differences in effective radii lead to large differences in attenuation coefficients, optical depth, ice water content and cloud radiative properties. For the Rain 10 (spheres) + C.6 (spheres) bimodal distribution, cloud reflectance varies from 2.6% for a cloud of 500 m in thickness to 16.9% for a cloud 3 km in thickness. Values of cloud absorptance are significantly larger than for the smaller particle modes alone, ranging from 7.6% for a cloud 500 m thick to 22.5% for a cloud 3 km thick. Much larger values of cloud reflectance and absorptance occur for the very large Rain 50 (spheres) + C.6 (spheres) distribution where values of cloud absorptance for this 3 km thick cloud are much larger than those values reported by Reynolds *et al.* The large-particle size distributions for which calculated values of cloud absorptance are in agreement with the measured values [i.e., the L_c = 175 (columns), L_c = 200 (columns) or Rain L (spheres)], also show relatively small values of reflectance (6.7–9.4%). Therefore, it may be concluded that the C.6 size distribution in combination with any of the large-particle modes cannot provide values of bulk cloud radiative characteristics similar to those reported by Reynolds *et al.*

TABLE 4.11. Percent cloud reflectance *(R)*, absorptance *(A)* and heating rates $\partial\theta/\partial t$ (°C h^{-1}) for various bimodal ice crystal size distributions. The large-particle size distributions (left-hand column) are combined with the spherical C.6 small-particle distribution to form various bimodal distributions. A second set of calculations are shown for the C.6 particle concentration increased by an order of magnitude. Computations are performed for a cloud with base height of 9 km and cloud thicknesses of 0.5, 1 and 3 km with solar zenith angle $\theta = 0°$.

Large-particle size distribution		Standard C.6 size distribution: Cloud thickness 500 m	1000 m	3000 m	Standard C.6 size distribution + large-particle concentration*10: Cloud thickness 500 m	1000 m	3000 m
Rain L (spheres)	R	1.4	2.9	9.4	5.9	12.2	31.1
	A	4.6	8.0	15.9	13.2	19.8	27.2
	$\frac{\partial\theta}{\partial t}$	1.03	0.93	0.69	2.95	2.29	1.18
Rain 10 (spheres)	R	2.6	5.5	16.9	18.8	32.0	49.9
	A	7.6	12.9	22.5	23.0	27.0	33.0
	$\frac{\partial\theta}{\partial t}$	1.71	1.49	0.98	5.14	3.14	1.43
Rain 50 (spheres)	R	5.6	11.6	29.7	36.1	46.3	55.2
	A	14.7	21.5	29.1	30.7	35.4	39.5
	$\frac{\partial\theta}{\partial t}$	3.30	2.49	1.27	6.88	4.11	1.72
Rain L (columns)	R	1.0	2.1	6.7	1.5	3.1	9.8
	A	3.4	5.9	11.5	4.1	7.1	14.0
	$\frac{\partial\theta}{\partial t}$	0.77	0.68	0.50	0.91	0.82	0.61
Rain 10 (columns)	R	1.1	2.3	7.4	2.6	5.4	16.7
	A	3.6	6.3	12.3	5.9	10.1	19.0
	$\frac{\partial\theta}{\partial t}$	0.81	0.73	0.53	1.32	1.17	0.83
Rain 50 (bullets)	R	1.3	2.6	8.3	4.1	8.5	24.4
	A	3.9	6.8	13.4	8.4	13.9	22.9
	$\frac{\partial\theta}{\partial t}$	0.88	0.79	0.58	1.88	1.61	0.99
Rain 50 (columns)	R	1.3	2.7	8.5	4.4	9.2	25.8
	A	4.0	6.9	13.6	8.9	14.5	23.3
	$\frac{\partial\theta}{\partial t}$	0.89	0.80	0.60	1.98	1.68	1.01
$L_c = 175$ (columns)	R	1.0	2.1	6.7	1.5	3.1	9.8
	A	3.4	5.9	11.4	4.0	7.0	13.7
	$\frac{\partial\theta}{\partial t}$	0.76	0.68	0.50	0.90	0.81	0.60
$L_c = 200$ (columns)	R	1.1	2.2	6.9	1.9	3.9	12.4
	A	3.5	6.0	11.7	4.5	7.9	15.4
	$\frac{\partial\theta}{\partial t}$	0.78	0.70	0.51	1.02	0.91	0.67

Table 4.11 also shows calculations in which the large-particle distributions have been increased in concentration by an order of magnitude. For such an increase in concentration of large particles, there are increases in the values of cloud reflectance and absorptance for all cloud thicknesses. For most of these bimodal size distributions the value of cloud reflectance remains smaller than that of cloud absorptance. The large-particle size distributions do provide increased values of cloud reflectance, but at the same time, produce much larger values of cloud absorptance. Here again cloud radiative properties comparable to those measured by Reynolds *et al.* cannot be obtained with any of the bimodal size distributions given in Table 4.11.

Comparison of Table 4.11 with Tables 4.9 and 4.10 show that the addition of the C.6 (spheres) small-particle distribution significantly affects the bulk radiative properties. Alone, the C.6 spheres monomodal distribution for a cloud thickness of 3 km have values of reflectance of 6.3% and cloud absorptance of 11.2%. The monomodal Rain L (columns), and the L_c = 175 or 200 (bullets and columns) distributions alone have corresponding values of reflectance of only 0.3% and absorptance values of about 7%. The corresponding bimodal distributions for the 3000 m thick cloud yield values of reflectance of 6.7% and absorptance of 11.5%. The radiative characteristics of these bimodal distributions are almost totally dominated by the small-particle C.6 (spheres) distribution.

A quite different situation occurs when the large-particle number density is increased by an order of magnitude. The C.6 (spheres) + Rain L (columns) bimodal distribution provides a value of reflectance of 9.8% and an absorptance of 14%. The value of reflectance is approximately the sum of the monomodal C.6 (spheres) and monomodal Rain L (columns) values. However, the bimodal value of absorptance is 14%, compared to 11.5% for the monomodal C.6 (spheres) and 10.3% for the Rain L (columns) shown in Table 4.9. Therefore, as expected, absorptance values are not additive. Similar behavior is observed for most of the other bimodal distributions.

As a next step in this sensitivity analysis, the small particle C.6 (spheres) number density was increased by an order of magnitude. In this case most bimodal distributions were again totally dominated by the effects of the small particles. Increasing both the C.6 (spheres) and large-particle distributions by an order of magnitude yielded only slightly larger values of reflectance (47–50%) and absorptance (24–25%) for most bimodal distributions.

TABLE 4.12. Cloud reflectance *(R)*, absorptance *(A)* and heating rates ($\partial\theta/\partial t$) (°C h^{-1}) for various ice crystal bimodal size distributions as discussed in the text, the cloud is 3000 m thick.

Large-particle size distribution		← Additional small- and large-particle distributions →			
		Nonspherical C.6 (bullets)	Nonspherical C.6 (bullets)*10	Nonspherical C.6 (bullets)*10 and large particles*10	C.5/10 + Large*10
Rain L (spheres)	*R*	4.3	15.5	35.9	47.2
	A	13.2	16.6	27.3	26.9
	$\frac{\partial\theta}{\partial t}$	0.57	0.72	1.19	1.16
Rain 10 (spheres)	*R*	13.2	22.9	52.3	55.6
	A	21.6	22.9	32.7	32.1
	$\frac{\partial\theta}{\partial t}$	0.94	1.00	1.42	1.39
Rain 50 (spheres)	*R*	28.9	34.3	56.6	56.9
	A	29.0	29.3	38.8	38.9
	$\frac{\partial\theta}{\partial t}$	1.26	1.27	1.68	1.69
Rain L (bullets)	*R*	1.3	12.8	15.5	41.3
	A	7.7	11.9	13.9	17.1
	$\frac{\partial\theta}{\partial t}$	0.34	0.52	0.61	0.74
Rain 10 (columns)	*R*	2.1	13.6	23.0	43.4
	A	8.8	12.8	19.5	20.7
	$\frac{\partial\theta}{\partial t}$	0.38	0.56	0.85	0.90
Rain 50 (bullets)	*R*	3.1	14.5	30.1	42.1
	A	10.2	14.0	23.1	17.7
	$\frac{\partial\theta}{\partial t}$	0.44	0.61	1.00	0.77
Rain 50 (columns)	*R*	3.3	14.7	31.4	46.4
	A	10.5	14.2	23.5	23.3
	$\frac{\partial\theta}{\partial t}$	0.46	0.62	1.02	1.01
L_c = 175 (bullets)	*R*	1.3	12.8	15.5	41.3
	A	7.7	11.8	13.6	16.8
	$\frac{\partial\theta}{\partial t}$	0.33	0.51	0.59	0.73
L_c = 200 (bullets)	*R*	1.5	12.9	17.3	42.1
	A	8.4	12.5	16.7	17.7
	$\frac{\partial\theta}{\partial t}$	0.36	0.54	0.73	0.77
L_c = 175 (columns)	*R*	1.4	12.8	16.2	41.6
	A	7.8	11.9	14.2	17.3
	$\frac{\partial\theta}{\partial t}$	0.34	0.52	0.62	0.75
L_c = 200 (columns)	*R*	1.7	13.1	18.9	42.4
	A	8.1	12.1	15.9	18.4
	$\frac{\partial\theta}{\partial t}$	0.35	0.53	0.69	0.80

The previous results were obtained using the C.6 (spheres) small-particle size distribution. Table 4.12 shows similar calculations but for different small-particle modes. Calculations using the nonspherical C.6 (bullets) small-particle mode provides values of cloud absorptance ranging between 8 and 13% for all but two of the bimodal size distributions; the very large particle spherical modes, as always, provide much larger values of absorptance. Smaller values of cloud reflectance result using the nonspherical C.6 (bullets) small-particle mode than with the spherical C.6 mode.

Next (Table 4.12), the small-particle concentration is increased by an order of magnitude. The values of cloud reflectance are smaller than those which resulted from the C.6 spherical, small-particle distribution. This behavior may be related to the smaller sized particles (effective radii) which result from the C.6 (bullets) distribution. The values of cloud absorptance in these two cases are nearly equivalent. Absorption, then, is dominated by the large particles. Increasing the concentration of small particles increases the value of cloud reflectance much faster than the value of cloud absorptance.

The next bimodal size distribution (Table 4.12) retains the nonspherical C.6 (bullets) concentration increased by an order of magnitude and also increases the large-particle mode concentration by an order of magnitude. Increasing the large-particle concentration increases values of cloud reflectance as well as cloud absorptance. However, for particles in the intermediate size range [with ice crystal mode lengths of 100–300 μm; i.e., $L_c = 175$, $L_c = 200$, Rain L (bullets)], the effect of increasing the large particle concentration by an order of magnitude only increases the values of cloud reflectance and absorptance by about 2–4%. Increasing large-particle concentrations has the largest effect upon cloud radiative properties for bimodal distributions with the largest particle sizes (i.e., Rain 10 and Rain 50). These results indicate that variations in large-particle size distributions (for ice particles in size ranges typically reported, i.e., $L_c \approx 200$ μm) do not have a significant impact upon cloud radiative characteristics. However, the concentration of very large ice crystals or hail may still have a significant effect upon cloud radiative properties (Sartor and Cannon, 1977) as shown using the Rain 10 and Rain 50 size distributions.

The C.5/10 small particle distribution (with 10 cm^{-3}) is next substituted for the nonspherical C.6 (bullets) distribution. The large-particle concentration is retained at a value an order of magnitude larger than "normal". Comparison of these results with those discussed previously shows that values of cloud reflectance and absorptance are not greatly influenced by the variation in the spherical, small particle size distribution. However, there is a large increase in the value of reflectance over the previous cases discussed in Table 4.12 for the large-particle columns and bullets. The C.5/10 distribution has a small-particle number density one order of magnitude larger than that given by the nonspherical C.6 (bullets). Decreasing the C.5 distribution by a second order of magnitude (to C.5/100) produced bimodal radiative properties similar to those of the nonspherical C.6 (bullets)*10. This once again demonstrates that particle number density, rather than particle size, is the dominant variable determining cloud albedo.

The C.5/10 + Rain L (bullets)*10 bimodal ice particle size distribution provides a cloud reflectance of 41.3%, with a cloud absorptance of 17.1%. Essentially equivalent results are obtained by replacing the Rain L (bullets) large-particle mode with either the Rain L (columns), $L_c = 175$ or $L_c = 200$ (bullets and columns) modes. This point is emphasized because it would appear that knowledge of the exact large-particle size distribution is unnecessary for flux calculations. An approximate function providing only the mode radius, width of the size spectrum and particle density is apparently sufficient. It is significant that the results do not depend upon a very precise specification of particle geometry.

It is also noted that these bimodal size distributions produce bulk cloud radiative characteristics which agree favorably with the measurements of Reynolds *et al.* and using the above arguments we deduce that it would then seem indicated that there were large numbers of small particles (~ 10 cm^{-3}) in the clouds measured by Reynolds *et al.* Retaining the C.5/10 small-particle distribution and decreasing the large-particle number density by an order of magnitude back to its original (normal $\approx 10^{-3}$ cm^{-3}) value leads to even better agreement with Reynolds *et al.* In these cases (not shown in Table 4.12) the value of reflectance decreases to about 38%, and the value of absorptance decreases to about 13–14%. The large-particle size distributions as measured by Griffith *et al.* (1979) (or approximated by the Rain 10 columns or bullets distribution) appears to provide a reasonable estimate of the large-particle sizes in such cirrus shields when complemented by a small-particle mode with a number density of about 10 cm^{-3}.

The results of Tables 4.11 and 4.12 show that the actual size distribution used to represent measured values of either the small-particle mode and/or the large-particle mode has little impact on the cloud radiative characteristics. These results indicate that if one knows the approximate large-particle sizes (to within about 50–100 μm) and concentrations (to within about a factor of 3), the cloud reflectance or absorptance will not vary by more than about 5% for a cloud $\lesssim 3$

km thick. These results suggest that actual size distributions need not be known exactly; furthermore, they indicate that nonspherical effects are likely to be "lost in the noise". It appears that a reasonable estimate of ice crystal lengths and length-to-width ratios, coupled with spherical Mie theory using effective sphere radii based upon equivalent areas, may provide realistic estimates of the radiative properties of cirrus clouds. Since the ice crystal concentrations, sizes and shapes are usually inadequately known and highly variable, detailed calculations using nonspherical corrections do not at present seem to be warranted. Perhaps of more significance is the fact that the presence of small particles has a dominant effect upon the radiative characteristics of cirrus clouds. These results indicate that microphysical observations should be as concerned with measuring the sizes and number densities of small particles as well as those of larger particles. While a detailed specification of the particle geometries of the small size particles is not essential, large-particle shapes may be more critical. Spherical particles such as hail may have significantly different radiative characteristics from those of ice crystals. The conclusions presented here are somewhat startling and, of course, need to be verified or modified on the basis of further calculations and measurements.

The results presented in Tables 4.11 and 4.12 were discussed as a function of the large-particle mode, for a constant small-particle mode. This procedure is reversed in Table 4.13 in which calculations are shown as a function of small-particle mode, for a constant large-particle mode. Cloud thickness is 3 km, except as noted, for a cloud base height of 9 km and solar zenith angle of $\theta = 0°$. The first large-particle mode considered in Table 4.13 is that of the $L_c = 200$ (bullets) distribution. The small-particle C.6 distributions, with small-particle concentrations of 1 cm^{-3}, result in very small values of cloud reflectance.

Calculations with the various C.5 (spherical, nonspherical, bullets and columns) small-particle distributions, with particle concentrations of 10^2 cm^{-3}, are considered next. These bimodal distributions with large small-particle concentrations yield values of cloud reflectance ranging between 50.3 and 72.9%, and values of cloud absorptances between 15 and 20%. Values of such magnitude are consistent with the measured values reported by Reynolds *et al.* Calculations using the small-particle nonspherical correction provide smaller values of cloud reflectance and much larger values of absorptance. These results suggest that the Chýlek resonance suppression theory cannot be used to describe the radiative characteristics of the clouds reported by Reynolds *et al.* since the absorptance values provided by the nonspherical theory appear to be much larger than measurements indicate. An explanation for this may be that these small particles have a preferred shape and orientation so that resonance suppression theory is not applicable. Another possibility is that resonance suppression theory, being an approximation, does not adequately describe the radiative characteristics of nonspherical particles found in cirrus clouds.

The calculations for the bimodal size distribution of C.5 (columns) + $L_c = 200$ (bullets) provide a cloud reflectance value of 50.3% and cloud absorptance of 15.7%, again similar to those reported by Reynolds *et al.* The cloud-averaged heating rate for this case is 0.68°C h^{-1}. Application of the actual cloud microphysical properties, cloud thickness and solar zenith angles should be able to provide even closer agreement with observations. However, such complete measurements are not currently available in the literature.

It is important to note that these results do provide reasonable agreement with measurements. Furthermore, these results argue for the presence of large concentrations of small particles. No combination of size distributions without large concentrations of small particles can reproduce the measurements of Reynolds *et al.* Of course, the small particle concentration of 10^2 cm^{-3} used in Table 4.13 may be too large by about an order of magnitude. Decreasing the small-particle number density by an order of magnitude led to a reduction in the value of reflectance by about 10–15% (to values of about 35–40%) with corresponding decreases in absorptance to about 13–14%. In any case the presence of significant quantities of small particles is indicated. The measurements of Reynolds *et al.* were taken in the cirrus shields of large cumulonimbus clouds. These highly convective systems may have rapidly transported supercooled cloud droplets to very high elevations where they may have been frozen rather than forming large crystals. Splintering of the large ice particles may also provide an explanation for the large numbers of small particles that these measurements suggest. In any case, the present results indicate that in cirrus clouds (with $T \lesssim -40°C$), the only way that large cloud reflectance values may occur is from the presence of significant quantities of smaller particles.

Comparison of Tables 4.8 and 4.13 shows that in the presence of large quantities of small particles (i.e., cirrus clouds with large cloud reflectance values), the cloud radiative characteristics are dominated by the small particles. The role of the large-particle mode is to decrease the value of cloud reflectance and increase that of cloud absorptance. For the case of the C.5 (columns), the presence of the $L_c = 200$ (bullets) results in a decrease in cloud reflectance from 58.7 to 50.3% and an increase in cloud absorptance from 14 to 15.7%.

The above results have been based upon the C.5

TABLE 4.13. Percent cloud reflectance (R), absorptance (A) and heating rates ($\partial\theta/\partial t$, °C h^{-1}) for various ice crystal bimodal size distributions as discussed in the text.

Small-particle size distribution		← Additional large-particle distributions →				
		L_c = 200 bullets Thickness 3 km	L_c = 200 columns Thickness 3 km	L_c = 200 columns*10 Thickness 3 km	L_c = 200 columns*10 Thickness 1 km	Rain 50 spheres Thickness 3 km
C.6 (spheres)	R	6.8	6.9	12.4	6.8	29.7
	A	11.9	11.7	15.4	8.0	29.1
	$\frac{\partial\theta}{\partial t}$	0.52	0.51	0.67	0.93	1.27
Nonspherical C.6 (bullets)	R	1.5	1.7	7.7	2.4	28.9
	A	8.4	8.1	13.0	6.6	29.0
	$\frac{\partial\theta}{\partial t}$	0.36	0.35	0.56	0.76	1.26
C.5 (spheres)	R	72.9	72.7	72.5	61.8	68.4
	A	20.5	20.5	20.8	18.7	26.2
	$\frac{\partial\theta}{\partial t}$	0.89	0.89	0.91	2.17	1.14
Nonspherical C.5	R	63.8	63.5	63.5	56.3	62.2
	A	32.1	32.2	32.3	27.3	33.9
	$\frac{\partial\theta}{\partial t}$	1.39	1.40	1.40	3.17	1.47
C.5 (bullets)	R	56.1	55.7	55.6	32.7	53.5
	A	16.7	16.1	18.5	11.6	28.3
	$\frac{\partial\theta}{\partial t}$	0.73	0.70	0.80	1.34	1.23
C.5 (columns)	R	50.3	49.8	50.1	26.7	50.0
	A	15.7	14.9	17.9	10.6	28.5
	$\frac{\partial\theta}{\partial t}$	0.68	0.65	0.78	1.23	1.24
Nonspherical C.5 (bullets)	R	31.0	30.4	31.2	22.1	34.1
	A	58.7	58.7	58.5	40.9	57.8
	$\frac{\partial\theta}{\partial t}$	2.55	2.55	2.54	4.75	2.51
Nonspherical nimbostratus top	R	70.5	70.3	70.2	67.1	68.8
	A	28.1	28.2	28.3	25.2	29.8
	$\frac{\partial\theta}{\partial t}$	1.22	1.23	1.23	2.92	1.29
Nimbostratus top (bullets)	R	66.6	66.4	66.1	48.3	61.7
	A	19.1	18.9	19.8	14.9	27.4
	$\frac{\partial\theta}{\partial t}$	0.83	0.82	0.86	1.73	1.20
Nimbostratus top (columns)	R	64.0	63.7	63.4	43.8	59.4
	A	18.7	18.4	19.6	14.0	27.7
	$\frac{\partial\theta}{\partial t}$	0.81	0.80	0.85	1.62	1.21

distribution. However, actual size distributions will have somewhat different values. The nimbostratus top (Chapter 2) size distribution has also been used to represent the small-particle mode. This distribution has larger particle sizes and an ice water content of about a factor of 3 greater than that for the C.5 distribution. These calculations show that the nimbostratus distribution provides larger values of cloud reflectance and absorptance than do similar C.5 distributions.

The next set of calculations shown in Table 4.13 substitute columns for bullets in the large-particle L_c = 200 mode. Somewhat smaller values of cloud reflectance and absorptance result from this assumption, but the variations are negligible. Increasing the large-particle concentration by an order of magnitude may be significant in some cases. Such an increase leads to significantly larger values of cloud reflectance and absorptance for the C.6-sized distributions which have a small (10^{-1} cm^{-3}) particle concentration. In the presence of a large concentration of small particles however, variations of the large-particle concentration are generally negligible. Nevertheless, an increase in the concentration of the large particles by an order of magnitude leads to an increase in cloud absorptance from 15 to 18% for the C.5 (columns) case; however, for the C.5 spherical case, variations in cloud absorptance are negligible.

The next set of calculations is for a cloud 1 km thick with the same large-particle distribution and concentration described above. Decreasing the cloud thickness by a factor of 3 decreases the value of both cloud reflectance and cloud absorptance, and increases the values of cloud-averaged heating rate since cloud heating is primarily located in the upper regions of the cloud. For the C.5 (columns) case this decrease in cloud thickness leads to a decrease in cloud reflectance from 50.1 to 26.7 and a decrease in cloud absorptance from 17.9 to 10.6%. Cloud-averaged heating rate has increased from 0.78 to 1.23°C h^{-1} for this case. Assuming bullets instead of columns in this 1 km thick cloud increases the value of cloud reflectance to 32.7%, absorptance to 11.6% and cloud-averaged heating rate to 1.34°C h^{-1}. Assuming spheres, much larger values are obtained. Even assuming the nimbostratus top distribution instead of the C.5 distribution for bullets and columns leads to larger values of both cloud reflectance and absorptance. Values of cloud reflectance vary from 26.7 to 61.8% and values of cloud absorptance from 10.6 to 18.7%, depending upon the assumed small-particle size distribution (C.5 or nimbostratus top, bullets or columns). These variations are much larger than similar variations in thicker clouds (3 km). Nevertheless, such values are consistent with the measurements of Reynolds *et al.*

As a final set of calculations the large-particle Rain 50 distribution, representing large particles such as hail, is used for a cloud 3 km thick. Very large values of cloud absorptance, ranging from 26.2 to 29.1%, result for the bimodal size distributions (excluding non-spherical cases). Cloud-averaged heating rates ranging from 1.14 to 1.27°C h^{-1} result from such cases. These results are similar to those obtained for large water drops.

4.3.4 Ice crystals embedded in water clouds

The results for water clouds are similar to those found for ice clouds. Calculations for very large-particle sizes (hail) provide very large values of cloud absorptance. Measurements by Sartor and Cannon indicate that ice crystals with concentrations of 0.4 cm^{-3} and ice water contents up to 7 g m^{-3} may be present in convectively active clouds. These extreme values of particle concentration and ice water content are larger than those modeled in the previous discussion. Table 4.14 models large ice crystals embedded in clouds with water droplets rather than small ice particles. Assuming the C.5 cloud droplet model with several droplet concentrations, various ice crystal models are used to represent this ice-water bimodal distribution. Cloud thickness is 3 km with cloud base at 3 km.

Calculations for ice crystals assuming the Rain L (columns), Rain 50 (columns) and Rain L (spheres) distributions provide similar values for cloud reflectance (77%) and absorptance (17%) for a concentration of 10^2 cm^{-3} of small droplets. For very large particles (Rain 50 spheres), cloud absorptance is 21.7%.

The measurements of Sartor and Cannon indicated that droplet concentrations are depleted rapidly in the presence of large ice particles. Decreasing the droplet concentration by a factor of 5 (C.5/5), to a value of 20 cm^{-3}, decreases the cloud reflectance to about 56.5% and increases cloud absorptance to about 18–19%. The cloud absorptance value of the Rain 50 (spheres) distribution is increased to 25.4%.

Decreasing the small-particle concentration by an additional factor of 4 (C.5/20), with a concentration of 5 cm^{-3}, decreases the cloud reflectance to 26–27% and decreases the cloud absorptance to 15–17%. Note that the value of cloud absorptance has remained nearly invariant while the value of cloud reflectance has decreased by approximately a factor of 3 for this decrease in droplet concentration by a factor of 20. This further decrease in droplet concentration leads to only a small increase in absorption for the large particle Rain 50 (spheres) distribution and a value of cloud reflectance of 42.1%.

For the Rain L and Rain 50 (columns) cases, a further decrease in the value of droplet concentration by a factor of 5 (C.5/100), with a droplet concentration of 1 cm^{-3}, decreases the cloud reflectance to values between 7.5 and 10% and decreases the cloud absorptance values to 12.5 and 15%. For these small concentrations of droplets, cloud radiative characteristics are determined primarily by the large particles. It is

TABLE 4.14. Percent cloud reflectance (*R*), transmittance (*T*) and absorptance (*A*) for bimodal size distributions consisting of water droplets (the C.5 distribution) with various droplet concentrations and ice crystals. Cloud thickness is 3 km, cloud base height 3 km, and solar zenith angle $\theta = 0°$.

Large-particle ice crystal size distribution		Droplet concentrations in a C.5 distribution C.5	C.5/5	C.5/20	C.5/100
Rain L (columns)	*R*	77.0	56.6	26.1	7.2
	T	5.8	25.3	58.3	80.0
	A	17.2	18.1	15.6	12.6
Rain 50 (columns)	*R*	76.9	56.6	27.6	9.6
	T	5.7	24.6	55.1	75.8
	A	17.4	18.8	17.3	14.6
Rain 50 (bullets)	*R*	76.1	56.5	32.1	17.0
	T	5.5	22.4	46.4	62.7
	A	18.4	21.1	21.5	20.3
Rain L (spheres)	*R*	76.9	56.6	26.9	8.6
	T	5.8	24.8	56.5	77.6
	A	17.3	18.6	16.6	13.8
Rain 50 (spheres)	*R*	73.5	56.9	42.1	34.1
	T	4.8	17.7	31.5	39.5
	A	21.7	25.4	26.4	26.4

identical to the procedure described in Davis *et al.* (1979a,b). The model computes absorption of solar radiation by water vapor in the regions above the cloud and within the cloud. The effects of scattering and absorption of solar radiation by water droplets are also included. All calculations are carried out for the same regions of the solar spectrum described in Chapter 1.

The Monte Carlo model is formulated in two phases. In the first phase, photon entry points are uniformly distributed over the top of a rectangular parallelepiped which is subdivided into smaller boxes representing either cloudy or neighboring clear regions. Conservative scattering of the photons is then simulated according to the Monte Carlo model described by McKee and Cox (1974), wherein the distance s_i between the ith $-$ 1 and the ith scattering events is given as the solution to

$$T_i = e^{-\tau_i} = \exp\left(-\Sigma\, \beta_j s_j\right)_i, \qquad (5.1)$$

where

$$s_i = \left(\Sigma\, s_j\right)_i.$$

In Eq. (5.1) T_i is the transmittance at the ith event which is randomly chosen between 0 and 1, τ_i the optical depth, β_j the volume extinction coefficient in the jth cloud box, and s_j the distance travelled by the photon in the jth box. The values of T_i and τ_i in Eq. (5.1) are average quantities with respect to wavelength since they result from utilization of a spectrally averaged value of β_j as explained below. Scattering from the direction of propagation (the z axis) is accomplished by a rotation of the coordinate system about the z axis by an angle γ, which is uniformly assigned a value between 0 and 2π. This is followed by a second rotation about the new or rotated x axis by an angle α, which is found as the solution to

$$PP(\alpha) = \left[\int_0^{\alpha} P(\alpha) \sin\alpha\, d\alpha\right] \Big/ \left[\int_0^{\pi} P(\alpha) \sin\alpha\, d\alpha\right], \qquad (5.2)$$

where $PP(\alpha)$ is randomly assigned a value between 0 and 1 and $P(\alpha)$ is the scattering phase function. As the scattering process continues the pertinent information of the photons' paths are recorded until they arrive at a cloud boundary. At this point, in the simulation of finite clouds, cloud exit coordinates and directions of propagation are recorded. In the simulation of infinite clouds, if the photon has arrived at other than the top or bottom face, it is re-entered at a point symmetric with respect to the plane which is parallel to the cloud face being exited and which passes through the center of the cloud.

The second phase of the model begins by simulating absorption occurring between the top of the atmosphere and the cloud top. Next, the recorded scattering information of phase 1 is recalled to calculate the absorptance within the cloud. Finally the absorptance by water vapor is calculated for regions outside the cloud due to scattering into the upward or downward hemispheres. Absorption due to water vapor is calculated according to the empirical fit of Howard *et al.* (1955), in a form modified by Liou and Sasamori (1975). The mean absorptance $\overline{A}$ in the band of width Δv (cm^{-1}) is given as

$$\overline{A} = [C + D \log_{10}(x + x_0)]/\Delta v, \qquad (5.3a)$$

where

$$x = \mu\, p_{\text{eff}}^{K/D}, \qquad (5.3b)$$

$$x_0 = 10^{-C/D}. \qquad (5.3c)$$

In this formulation K, C and D are empirically derived constants, μ is the optical depth due to water vapor and P_{eff} the effective pressure. Numerical values of the above constants are tabulated by Liou and Sasamori (1975). Absorption by water droplets is calculated in Phase 2 by a simple weighting scheme according to

$$E_1 = E_0\, \omega, \qquad (5.4)$$

where ω is the single-particle scattering albedo and E_0 and E_1 are the photon's energy before and after scattering, respectively. We point out here that each photon in the simulation process is actually a statistical representation of a packet of many photons, therefore we do not imply that Eq. (5.4) applies to a single photon. As a photon packet traverses the atmosphere and cloud, its energy is successively reduced by water vapor and droplet absorptions so that it enters each new interaction with the correct initial energy.

Several of the quantities in Eqs. (5.1)–(5.4) are wavelength-dependent, namely, β_j, $P(\alpha)$, K, C, D, Δv and ω. An exact result requires a calculation at all wavelengths between 0.3 and 8.0 μm. Computer time considerations for such a calculation are prohibitive; therefore, the following approximation was made. The 0.3–8.0 μm spectrum was divided into three regions following the method used by Welch *et al.* (1976). Region I, for wavelengths from 0.8–2.7 μm is characterized by strong vapor absorption, while in region II, comprising the remainder of the near-IR spectrum, droplet absorption is predominant. Region III comprises the remainder of the solar spectrum. A single volume extinction coefficient and phase function were assigned to each region by weighting several volume extinction coefficients and several sets of phase function expansion coefficients according to the magnitude of the extraterrestrial solar intensity characteristic of the various wavelengths within the respective regions. Numerical data pertinent to this averaging process for the various droplet distributions are given in Chapter 2. The averaged phase function and extinction coefficients of regions I, II and III are used separately in Phase 1 of the model to generate three sets of photon paths. Phase 2 utilizes the generated photon paths for each region separately, applying the remainder of the wavelength-dependent parameters band by band, as if each set of photon paths from Phase 1 represented the actual paths of photons in the individual bands in question. In this manner, a calculation for the entire solar spectrum is accomplished in three model runs.

5.4 *Calculations for finite clouds*

Most previous reports of the radiation fields associated with finite clouds have been concerned with conservative scattering. Davis *et al.* (1979a) and Davies (1978) have included droplet and/or water vapor absorption within the cloud. Calculations by Davis *et al.* consider only two sizes of cubic clouds. The present computations will give values for the radiation field for a variety of cloud geometries.

5.4.1 The effect of cloud geometry upon the radiation field

Values of bulk cloud reflectance (R), transmittance (T) and absorptance (A) for a variety of cloud geometries are given in Table 5.1. The solar zenith angle $\theta = 0°$. The Best (1951) drop size distribution with a liquid water content of $w_L = 0.1$ g m^{-3} has been used. Attenuation coefficients for this size distribution were given in Chapter 2. Values of Δx, Δy and Δz (km) define the cloud geometry. Bulk cloud reflectance, transmittance and absorptance are defined in Chapter 2. Column A refers to these values calculated in terms of the energy incident at cloud top. Column B refers to these values calculated in terms of the energy incident at the top of the earth's atmosphere. For instance, reflectance in column A is defined as the ratio of energy reflected upward through the top face of the cloud to the incident energy transmitted downward through this same face. Transmittance is defined as the ratio of energy transmitted through the bottom face of the cloud

Table 5.1. Percent cloud reflectance (R), transmittance (T) and absorptance (A) for a variety of cloud geometries at solar zenith angle $\theta = 0°$ as discussed in the text. For a detailed explanation of the normalization used in columns A, B and C see p. 70.

Case	Δx (km)	Δy (km)	Δz (km)	Cloud top (km)		A Top of cloud	B Top of atmosphere	C Remote sensing (DR)
1	0.21	0.21	0.46	1.45	*R*	11.9	9.2	22.9
					T	2.2	1.7	
					A	1.8	1.4	
2	0.42	0.42	0.46	1.45	*R*	19.6	15.2	29.2
					T	7.8	6.1	
					A	2.5	1.9	
3	0.84	0.84	0.46	1.45	*R*	30.4	23.7	34.7
					T	16.8	13.0	
					A	3.3	2.6	
4	0.21	0.21	0.96	1.95	*R*	11.9	9.4	23.5
					T	0.01	0.01	
					A	2.0	1.6	
5	0.42	0.42	0.96	1.95	*R*	19.6	15.5	31.9
					T	0.3	0.2	
					A	2.9	2.3	
6	0.84	0.84	0.96	1.95	*R*	30.9	24.5	39.5
					T	2.1	1.7	
					A	4.1	3.3	
7	1.68	1.68	0.96	1.95	*R*	44.4	35.0	45.6
					T	7.1	5.6	
					A	5.7	4.5	
8	3.36	0.42	0.96	1.95	*R*	27.2	21.5	37.7
					T	1.2	0.9	
					A	3.7	3.0	
9	0.84	0.84	2.02	3.01	*R*	30.8	25.0	41.2
					T	0.03	0.03	
					A	5.4	4.4	
10	0.42	0.42	0.89	9.0	*R*	21.4	21.0	42.8
					T	0.3	0.2	
					A	6.2	6.1	

to the incident cloud top energy. Due to absorption by water vapor above the cloud in the longer wavelength regions, the incident energy at cloud top is smaller than that at the top of the atmosphere. Therefore, the bulk cloud properties referenced to the top of the atmosphere (column B) show smaller values than those referenced to cloud top (column A). For finite clouds a large portion of the total energy incident at cloud top exits the cloud sides. Some of this energy exits the cloud in the upward hemisphere and is transmitted back out of the atmosphere. A portion of the energy in the upper hemisphere is absorbed by water vapor before reaching the top of the atmosphere. Column C gives the reflectance of the cloud at the top of the atmosphere, defined as the ratio of total energy reflected out of the atmosphere by the cloud, (through both the top and side faces) to incident energy measured at the top of the atmosphere, hereafter referred to as directional reflectance *(DR)*.

The first case considered in Table 5.1 is for a cloud of horizontal dimensions 0.21 km on a side, 0.46 km thick, with cloud top height of 1.45 km. The thickness-to-width ratio for this cloud is about 2/1. In terms of energy incident at cloud top, 11.9% is reflected from the top of the cloud, 2.2% is transmitted through the bottom of the cloud, and 1.8% is absorbed by droplets and water vapor within the cloud. Therefore, 84.1% of the total energy incident at cloud top exits through the sides. In terms of incident energy at the top of the atmosphere, 9.2% is reflected from cloud top, 1.7% transmitted through the bottom of the cloud and 1.4% absorbed within the cloud. The cloud reflects back out of the top of the atmosphere 22.9% of the incident energy at the top of the atmospheric column containing the cloud; this energy can be remotely sensed by satellites. However, an aircraft flying directly above the cloud would measure a value of only about 12%.

Doubling the width of the cloud (t/w = 1/1) while keeping the cloud thickness unchanged (Case 2), sharply increases the value of cloud reflectance, from a value of 11.9% to a value of 19.6%. Transmittance increases from 2.2 to 7.8% and absorptance from 1.8 to 2.5%, the total energy reflected out of the earth's atmosphere increases to 29.2%.

Doubling the width of the cloud again (t/w = 1/2) while keeping the cloud thickness unchanged (Case 3) increases reflectance to 30.4%, transmittance to 16.8% and absorptance to 3.3%. In this case only 49.5% of the incident energy at cloud top exits the sides, while 34.7% is reflected out of the earth's atmosphere.

Cases 1–3 show that cloud reflectance rapidly increases with the horizontal size of the cloud. However, the increase in cloud transmittance is even more striking. A very "tall" cloud with a large thickness-to-width ratio has very little energy transmitted through cloud base, but most of the energy is transmitted through the cloud sides. As noted by Davies (1978) this effect is probably responsible for the extreme darkness observed beneath some thunderstorm clouds. A similar situation is reported by McKee and Klehr (1976) who showed that cloud towers should appear dark against the lower cloud field background as viewed from a satellite. The small turrets or towers reflect a much smaller portion of the total incident solar radiation than a cloud with a smaller thickness-to-width ratio.

Cases 4–8 consider a cloud 0.96 km in thickness, with a cloud top height of 1.95 km for various cloud widths. Case 4 considers a cloud 0.21 km in width. The values of bulk cloud radiative characteristics are similar to those for the cloud with half the thickness (Case 1). Cloud reflectance and absorptance are not significantly different for these two cloud thicknesses. However, the thicker cloud (Case 4) has a significantly smaller value of cloud transmittance (0.01%). For this cloud with a thickness-to-width ratio of about 4/1, essentially no energy is transmitted through cloud base, and 86.1% is transmitted through the cloud sides.

Doubling ($t/w \approx 2/1$) and quadrupling ($t/w \approx 1/1$) the width of the cloud sharply increases the cloud reflectance (see Cases 5 and 6). However, the values of cloud reflectance for the thicker (0.96 km) cloud (Cases 4–6) are almost identical to those of the thinner (0.46 km) cloud (Cases 1–3). Cloud reflectance measured at cloud top appears to be a function of horizontal cloud dimension to a far greater extent than to cloud thickness for these small finite clouds. However, the thicker cloud reflects a larger amount of total energy back through the top of the atmosphere (DR = 39.5% in Case 6 compared to 34.7% in Case 3). The value of cloud absorptance increases both with increasing cloud thickness and with increasing horizontal extent. Cloud transmittance remains lower than that for the thinner cloud. For the cloud with $t/w \approx 1$ (Case 6), 62.9% of the energy exits the sides. The thinner cloud with the same t/w ratio (Case 2) had 70% of the energy exiting the sides. Therefore, as the cloud thickness increases, keeping the same t/w ratio, a greater portion of the energy is reflected out of the top of the cloud, a smaller portion exits the sides and a smaller portion exits the base of the cloud; cloud absorptance increases slightly.

Doubling the cloud diameter once again (Case 7) to a t/w ratio of 1/2 increases cloud reflectance to 44.4%, transmittance to 7.1% and absorptance to 5.7%. Approximately 43% of the incident energy exits the sides and 45.6% of the incident energy is reflected back out of the atmosphere. In this case, the majority of the reflected energy exits the cloud top rather than the cloud sides.

The results of the previous calculations are for clouds

with square horizontal areas. Case 8 considers a cloud with one side very much longer than the other (3.36 km × 0.42 km). The horizontal area of this cloud is intermediate in value between that of the 0.84 km square cloud and that of the 1.68 km square cloud. However, the bulk cloud properties are smaller than those of the 0.84 km square cloud (Case 6). Therefore, the cloud radiative characteristics are a function of cloud geometry and not simply a function of horizontal cloud area. The cloud radiative characteristics (R, T and A) increase as one side of the cloud is lengthened (Cases 5 and 8). However, this increase in the values of R, T and A is smaller for the elongated cloud than for a square cloud of equal area.

Case 9 considers a cloud 0.84 km square and 2 km in thickness. Comparison of Cases 3 and 6 with Case 9 shows that the value of reflectance remains essentially invariant as cloud thickness increases while keeping cloud horizontal extent fixed. However, as cloud thickness increases, cloud absorptance increases and cloud transmittance decreases.

For a solar zenith angle of $\theta = 0°$, Case 10 considers a cloud of 0.42 km square and 0.89 km thickness with its top at 9 km. This cloud is slightly thinner than the cloud described by Case 5 (0.96 km thick), but has a larger value of reflectance and more than twice the value of absorptance; the value of transmittance remains unchanged. As in the infinite cloud cases discussed in Chapters 2 and 3, when the cloud is raised there is a corresponding increase in absorbed energy. As base height is raised from about 1 km (Case 5) to about 8 km (Case 10) approximately twice as much energy is absorbed within the cloud. Coupling this increased energy absorption (of about a factor of 2) with a decrease in air density of about a factor of 3, leads to increased heating rates of about a factor of 6 between low and high clouds, both for finite and plane-parallel cases. Of particular interest is the fact that the total energy reflected back out of the earth's atmosphere has increased from about 32% for the low-lying cloud (Case 5) to about 43% for the upper-level cloud (Case 10). The difference is primarily due to water vapor absorption above the cloud. Photons exiting the lower-level cloud have a high probability of being absorbed by the water vapor before leaving the earth's atmosphere.

Table 5.2 shows similar calculations for a solar zenith angle of $\theta = 30°$, and Table 5.3 for a solar zenith angle of $\theta = 60°$. Energy incident at cloud top is proportional to $\Delta x \cos \theta$ while energy incident on the cloud

TABLE 5.2. Percent cloud reflectance (R), transmittance (T) and absorptance (A) for a variety of cloud geometries at solar zenith angle $\theta = 30°$ as discussed in the text. For a detailed explanation of the normalization used in columns A, B and C see p. 73.

Case	Δx (km)	Δy (km)	Δz (km)	Cloud top (km)		A Top of cloud	B Top of atmosphere	C Remote sensing (DR)
1	0.21	0.21	0.46	1.45	R	6.0	4.6	18.5
					T	17.5	13.5	
					A	1.3	1.0	
2	0.42	0.42	0.46	1.45	R	14.6	11.2	25.3
					T	17.9	13.7	
					A	1.9	1.5	
3	0.84	0.84	0.46	1.45	R	26.6	20.5	31.4
					T	21.0	16.2	
					A	2.7	2.1	
4	0.21	0.21	0.96	1.95	R	4.0	3.1	21.0
					T	10.6	8.3	
					A	1.6	1.2	
5	0.42	0.42	0.96	1.95	R	10.2	8.0	26.6
					T	10.6	8.3	
					A	2.3	1.8	
6	0.84	0.84	0.96	1.95	R	22.1	17.2	33.8
					T	10.4	8.1	
					A	3.3	2.6	
7	1.68	1.68	0.96	1.95	R	37.1	29.0	40.8
					T	11.3	8.8	
					A	4.6	3.6	
8	0.42	3.36	0.96	1.95	R	26.5	20.7	36.6
					T	3.7	2.9	
					A	3.2	2.5	
9	3.36	0.42	0.96	1.95	R	14.8	11.6	30.1
					T	13.5	10.6	
					A	3.0	2.3	

TABLE 5.3. Percent cloud reflectance *(R)*, transmittance *(T)* and absorptance *(A)* for a variety of cloud geometries at solar zenith angle $\theta = 60°$ as discussed in the text. For a detailed explanation of the normalization used in columns A, B and C see below.

Case	Δx (km)	Δy (km)	Δz (km)	Cloud top (km)		A Top of cloud	B Top of atmosphere	C Remote sensing (DR)
1	0.21	0.21	0.46	1.45	*R*	5.6	4.2	22.6
					T	15.9	11.8	
					A	0.8	0.6	
2	0.42	0.42	0.46	1.45	*R*	14.2	10.6	26.9
					T	20.1	15.0	
					A	1.2	0.9	
3	0.21	0.21	0.96	1.95	*R*	3.0	2.3	24.2
					T	8.6	6.5	
					A	1.0	0.8	
4	0.42	0.42	0.96	1.95	*R*	8.3	6.3	28.9
					T	11.8	8.9	
					A	1.6	1.2	
5	0.84	0.84	0.96	1.95	*R*	19.3	14.5	33.3
					T	13.9	10.5	
					A	2.3	1.7	
6	1.68	1.68	0.96	1.95	*R*	33.9	25.6	39.2
					T	15.1	11.4	
					A	3.0	2.3	
7	0.21	0.21	2.02	3.01	*R*	1.7	1.3	26.3
					T	5.5	4.3	
					A	1.3	1.0	
8	0.84	0.42	0.96	1.95	*R*	10.4	7.9	29.4
					T	14.2	10.7	
					A	1.9	1.4	
9	0.42	0.84	0.96	1.95	*R*	14.6	11.0	31.7
					T	10.8	8.1	
					A	1.8	1.4	
10	3.36	0.42	0.96	1.95	*R*	12.5	9.4	30.6
					T	16.0	12.1	
					A	2.2	1.7	
11	0.42	3.36	0.96	1.95	*R*	28.4	21.5	37.9
					T	5.8	4.3	
					A	2.0	1.5	

sides is proportional to $\Delta z \sin \theta$; the total energy incident upon the cloud is then proportional to $\Delta x \cos \theta + \Delta z \sin \theta$. Column A shows values of reflectance *(R)*, transmittance *(T)* and absorptance *(A)* calculated in terms of the total energy incident upon the cloud (top and sides). Column B provides bulk radiative properties calculated relative to the energy incident at the top of the atmosphere, and includes incident radiation through the cloud sides as well as the cloud top, however, only energy emerging through the top and bottom faces are used to compute reflectance and transmittance respectively. Column C gives the ratio of energy scattered by the cloud top and sides out of the earth's atmosphere to the extraterrestrial incident cloud energy (top plus sides).

In order to appreciate the effect of solar zenith angle upon the finite cloud bulk radiative characteristics, Tables 5.1, 5.2 and 5.3 should be consulted together. Case 1 considers a 0.21 km square cloud, 0.46 km thick, with cloud top at 1.45 km. The value of cloud reflectance (the ratio of energy transmitted out of the top of the cloud to total energy incident upon the cloud * 100) decreases rapidly with increasing solar zenith angle. The value of 11.9% at $\theta = 0°$ has decreased to 6% at $\theta = 30°$ and to 5.6% at $\theta = 60°$. However, the value of transmittance has increased sharply from 2.2% at $\theta = 0°$ to 17.5% at $\theta = 30°$ and to 15.9% at $\theta = 60°$. The large increase in the fraction of incident energy (on the cloud top plus sides) which is transmitted through cloud base at the larger zenith angles is due to the reduced optical path lengths for photons incident on the illuminated side face near the lower edge of the cloud. This edge effect is greater for a solar zenith angle of 30° than for the 60° zenith angle case as indicated by the transmittance values. These edge effects are also greatest for a "tall" cloud with a large thickness-to-width ratio and decrease with decreasing value of *t/w*. The value of absorptance decreases with increasing solar zenith angle when referenced to total energy incident upon the cloud. In terms of energy incident at the top of the atmosphere (Column B), the values of R, T and A are decreased due to water vapor absorption above the cloud. In terms of radiation reflected back out of the earth's atmosphere Column C indicates that about 23% of the incident energy is reflected back to space for a solar zenith angle of $\theta = 0°$; for $\theta = 30°$ this amount decreases to 18.5%, and increases to 22.6% for $\theta = 60°$. However, in terms of total energy, the clouds with $\theta = 30°$ and $\theta = 60°$ intercept 1.37 times more energy than at $\theta = 0°$. Assuming a solar constant value of 1368 W m^{-2}, the total amount of energy reflected back out of the earth's atmosphere increases with solar zenith angle to values of 313.3, 345.7 and 422.3 W m^{-2}, respectively, for solar zenith angles of $\theta = 0°$, 30° and 60°.

Doubling the cloud width (Case 2 in Tables 5.1, 5.2 and 5.3) increases the values of R, T and A at all solar zenith angles. The relative increase in the value of reflectance with increasing cloud width is greatest at the larger zenith angles. The increase is from 11.9 to 19.6% at $\theta = 0°$, 6.0 to 14.6% at $\theta = 30°$ and 5.6 to 14.2% at $\theta = 60°$. Similar increases are observed for a further doubling of cloud width (Case 3 of Tables 5.1 and 5.2).

At a solar zenith angle of $\theta = 0°$, increasing cloud depth (Cases 1 and 4; and 3, 6 and 9 of Table 5.1) has little effect upon the value of cloud reflectance. Cloud transmittance decreases sharply and cloud absorptance increases slightly in these cases.

However, for the same cloud width, an increase in cloud thickness for $\theta > 0°$ leads to decreased values of cloud reflectance and transmittance and a slight increase in the values of absorptance (Cases 1, 3 and 7; and 2 and 4 of Table 5.3). These variations are largest for the largest solar zenith angles. As the value of t/w increases, less of the total incident radiation exits either the top or bottom cloud faces, so that the values of R and T decrease. The value of cloud absorptance increases with increasing cloud thickness or width, but decreases with increasing solar zenith angle when referenced to total cloud incident energy.

Cases 8 and 9 of Table 5.2 and Cases 8–11 of Table 5.3 consider the radiative characteristics of an elongated cloud. The solar beam illuminates the x-z face for $\theta > 0°$. Case 8 of Table 5.2 considers an elongated cloud (0.42 km × 3.36 km) oriented so that the shorter side and top of the cloud receive direct illumination. A cloud reflectance of 26.5% results in this case, considerably smaller than the value of 37.1% obtained for the 1.68 km square cloud. The values of T and A are also considerably smaller than for the square cloud. The elongated cloud "leaks" photons from the long side, thereby decreasing the values of R, T and A. However, the elongated cloud has a greater value of reflectance than does the 0.84 km square cloud, in contrast to the results for $\theta = 0°$ (Table 5.1). Nevertheless, the elongated cloud has a significantly smaller value of transmittance than does the square cloud, even though the elongated cloud has twice the area.

Case 9 of Table 5.2 considers this same elongated cloud oriented so that its longer side and top receive direct illumination. This change in orientation decreases the value of reflectance by about a factor of 2, but increases the value of transmittance by more than a factor of 3. The value of cloud absorptance is only slightly affected. This elongated cloud has only slightly larger values of R, T and A than does the 0.42 km square cloud, even though the elongated cloud has eight times the area of the square cloud.

It is also seen from comparing Cases 8 and 9 that the illuminated short side case (Case 8) scatters a greater fraction of the incident energy back to the top of the atmosphere than does the illuminated long side case. This is principally a result of the lower edge effect becoming more dominant in Case 9.

Similar behavior is noted at larger zenith angles (Table 5.3). Cases 8 and 9 consider an elongated cloud with one side twice as long as the other (0.42 km × 0.84 km). The values of R and A are intermediate in value between those of the 0.42 km square cloud and the 0.84 km square cloud. Orienting the elongated cloud so that its smaller side is illuminated increases the value of R and decreases the value of T over those values obtained with the larger side illuminated. These differences become more significant as the ratio of sides increases (Cases 10 and 11). Top of the cloud reflectance values (column A) are nearly twice as large for the highly elongated cloud (ratio of sides of 8/1) oriented with the small side illuminated. However, this same cloud reflects back to space (column C) only 37.9% of the incident energy, compared to 30.6% for the cloud oriented so that its long side is illuminated; the latter values include radiance components from the cloud sides as well as the cloud top.

5.4.2 Monomodal drop size distributions

The previous discussion of the effect of cloud geometry upon the cloud radiative properties has utilized the Best drop size distribution for a liquid water content of $w_L = 0.1$ g m^{-3}. However, the results of Chapters 2, 3 and 4 for plane-parallel clouds indicate that the choice of drop size distribution strongly influences cloud radiative characteristics. In contrast, the results of Reynolds *et al.* (1978) and McKee and Klehr (1977), using the C.1 and C.3 drop size distributions for visible radiation alone, indicate that the choice of drop size distribution may not be an important consideration in finite clouds. These conclusions will be examined in more detail in this section.

Attenuation coefficients and phase function expansion coefficients for the various drop size distributions discussed below were given in Chapters 2 and 3. For the purposes of comparison, a cloud with horizontal dimensions of 0.84 km × 0.84 km was chosen. In most cases, cloud thickness was chosen to be 0.96 km with cloud top height of 1.95 km and solar zenith angle of $\theta = 0°$.

The results of the first six drop size distributions in Table 5.4 are for nimbostratus, stratocumulus and stratus clouds, with values representative of cloud bases as well as cloud tops. The values of cloud reflectance for this nearly cubic cloud are significantly larger for drop size distributions representative of cloud tops than those values representative of cloud bases. Values of cloud reflectance vary between 20.8 and 46%, solely as a function of the drop size distribution. The magnitude of these variations are equivalent to a variation of thickness-to-width *(t/w)* ratios ranging from 2/1 to 1/2 (Cases 5–7, Table 5.1). Therefore, for all solar wavelength reflectance values, one must conclude that the choice of drop size distribution may be of equal importance to that of cloud geometry. The total radiation scattered back out of the earth's atmosphere is also shown to be strongly dependent upon the drop size distribution, with variations between 32.5 and 47.3%.

For plane-parallel clouds, absorptance was nearly invariant to the choice of drop size distribution. However, for finite clouds, this is definitely not the case. Absorptance varies from about 4% to 9%, with the largest values associated with the largest values of cloud reflectance (drop size distributions representative of cloud tops). The amount of radiation scattered

TABLE 5.4. Percent cloud reflectance *(R)*, transmittance *(T)* and absorptance *(A)* for a variety of monomodal drop size distributions. For a detailed explanation of the normalization used in columns A, B and C see p. 70 for $\theta = 0°$ and p. 73 for $\theta > 0°$.

Droplet distribution	Zenith angle (deg)	Δx (km)	Δy (km)	Δz (km)	Cloud top (km)		A Top of cloud	B Top of atmosphere	C Remote sensing (DR)
Nimbostratus top	0	0.84	0.84	0.96	1.95	*R*	45.8	36.1	47.3
						T	0.2	0.2	
						A	12.1	9.5	
Nimbostratus base	0	0.84	0.84	0.96	1.95	*R*	29.1	22.9	38.7
						T	1.8	1.4	
						A	5.4	4.3	
Stratocumulus top	0	0.84	0.84	0.96	1.95	*R*	43.9	34.6	46.5
						T	0.4	0.4	
						A	9.4	7.4	
Stratocumulus base	0	0.84	0.84	0.96	1.95	*R*	23.3	18.4	34.5
						T	3.0	2.3	
						A	4.8	3.8	
Stratus top	0	0.84	0.84	0.96	1.95	*R*	35.2	27.8	42.8
						T	1.3	1.0	
						A	4.7	3.7	
Stratus base	0	0.84	0.84	0.96	1.95	*R*	20.8	16.4	32.5
						T	3.9	3.1	
						A	4.5	3.5	
Nimbostratus top	60	0.84	0.84	0.96	1.95	*R*	23.5	17.7	35.8
						T	7.8	5.9	
						A	9.3	7.0	
Nimbostratus base	60	0.84	0.84	0.96	1.95	*R*	19.3	14.5	32.4
						T	13.8	10.4	
						A	3.3	2.5	
C.5	0	0.84	0.84	0.96	1.95	*R*	31.2	24.9	40.1
						T	1.6	1.3	
						A	5.8	4.6	
Rain 10	0	0.84	0.84	0.96	1.95	*R*	0.9	0.8	2.5
						T	70.2	55.4	
						A	6.3	5.0	
Rain 50	0	0.84	0.84	0.96	1.95	*R*	2.9	2.3	7.6
						T	40.4	31.8	
						A	10.2	8.1	
Rain 50	0	0.84	0.84	2.02	3.01	*R*	2.9	2.4	10.7
						T	8.4	6.9	
						A	13.8	11.2	

out the cloud sides varies from 46% (stratocumulus top) to 69% (stratocumulus base), a difference of 23%. Therefore, the droplet size distributions which are representative of cloud tops 1) scatter more radiation out of the cloud tops; 2) scatter more radiation back out of the atmosphere; 3) absorb more radiation; 4) transmit less radiation out the cloud bases; and 5) "leak" less radiation out the cloud sides than do droplet size distributions which are representative of cloud bases.

Calculations are also shown for the nimbostratus (top and base) droplet size distributions at $\theta = 60°$. The distributions representative of cloud tops scatter more radiation out of the cloud tops (and out of the atmosphere) and absorb more radiation while transmitting less energy through the cloud base. Values of bulk radiative properties for the C.5 drop size distribution are similar to those obtained with the stratus top and nimbostratus base distributions.

Calculations for the large-drop, Rain 10 and Rain 50 distributions are also given in Table 5.4. For the 0.96 km thick clouds these large-drop distributions scatter only about 1–3% of the incident radiation out the cloud top and about 2.5–7.6% out of the atmosphere; 6–10% is absorbed, about 70% transmitted out the cloud base and about 23% "leaked" out of the cloud sides. Increasing the cloud thickness of the Rain 50 distribution to 2.02 km results in virtually no change in the amount of radiation leaving the cloud. However, the radiation scattered out of the atmosphere is increased substantially for this case, with 13.8% absorbed, 8.4% transmitted out the cloud base and 74.9% "leaking" out the cloud sides. Clearly, radiative differences exist between clouds with large and small droplet distributions in both the finite and the infinite cases.

Calculations by Reynolds *et al.* (1978) and McKee and Klehr (1977) were for a wavelength of 0.70 μm. However, the calculations reported in Table 5.4 include the entire solar spectrum and include droplet absorptance. Therefore, the conclusions, that variations in cloud brightness due to changes in microphysical structure are small, only apply to the short-wavelength (visible) portion of the spectrum (< 0.7 μm) and not to the solar spectrum as a whole.

5.4.3 Bimodal drop size distributions

The previous results were obtained for monomodal drop size distributions. However, the results of Chapter 3 for bimodal drop size distributions in plane-parallel clouds showed that the presence of large drops significantly affected the cloud radiative characteristics. In such cases cloud reflectance decreased and cloud absorptance increased. The cloud radiative properties were shown to be particularly sensitive to the concentration of small droplets.

TABLE 5.5. Percent cloud reflectance (*R*), transmittance (*T*) and absorptance (*A*) for a variety of bimodal drop size distributions and cloud geometries. For a detailed explanation of the normalization used in columns A, B and C see p. 70.

Droplet distribution	Zenith angle (deg)	Δx (km)	Δy (km)	Δz (km)	Cloud top (km)		A Top of cloud	B Top of atmosphere	C Remote sensing (DR)
C.5 + Rain 50	0	0.84	0.84	0.96	1.95	*R*	31.7	25.0	40.0
						T	1.4	1.1	
						A	8.9	7.0	
C.5/10 + Rain 50	0	0.84	0.84	0.96	1.95	*R*	6.0	4.8	14.6
						T	19.7	15.5	
						A	9.6	7.6	
C.5/100 + Rain 50	0	0.84	0.84	0.96	1.95	*R*	3.1	2.4	8.1
						T	38.1	30.1	
						A	10.1	8.0	
C.5/10 + Rain 50	0	0.42	0.42	0.96	1.95	*R*	3.3	2.6	11.2
						T	9.1	7.1	
						A	7.9	6.3	
C.5/10 + Rain 50	0	1.68	1.68	0.96	1.95	*R*	11.7	9.2	18.2
						T	31.6	24.9	
						A	11.1	8.8	
C.5/10 + Rain 50	0	0.42	0.42	2.02	3.01	*R*	3.3	2.7	12.0
						T	0.2	0.2	
						A	9.5	7.7	
C.5/10 + Rain 50	0	0.84	0.84	2.02	3.01	*R*	6.5	5.3	17.9
						T	1.4	1.1	
						A	11.8	9.6	
C.5/10 + Rain 50	0	0.84	0.84	3.21	4.20	*R*	6.3	5.4	18.6
						T	0.06	0.05	
						A	14.0	11.8	

Table 5.5 shows similar calculations of cloud reflectance *(R)*, transmittance *(T)* and absorptance *(A)* for various cloud geometries and bimodal drop size distributions at a solar zenith angle of $\theta = 0°$. The bimodal drop size distribution consisting of the C.5 small-particle mode in conjunction with the Rain 50 large-particle mode is considered first. A cloud 0.84 km square with cloud thickness of 0.96 km and cloud top at 1.95 km is used as the basis for comparison with results obtained with monomodal drop size distributions.

Referenced to the incident energy at cloud top (column A), the C.5 + Rain 50 bimodal distribution provides a reflectance of 31.7% and absorptance of 8.9%. The C.5 monomodal drop size distribution provides a value of reflectance of 31.2% and absorptance of 5.8% (Table 5.4). The addition of the large drops, therefore, has not appreciably altered the cloud reflectance or total energy scattered out of the atmosphere. However, the presence of the large drops has significantly increased the cloud absorptance. The value of reflectance for the monomodal Rain 50 distribution is only 2.9%, with an absorptance of 10.2%. The bimodal distribution has a smaller value of absorptance than that of the large-drop distribution alone, which is consistent with similar findings for plane-parallel clouds. Only 58% of the incident energy is lost from the cloud sides for the bimodal distribution, compared to 62.3% for the C.5 monomodal distribution. Therefore, the primary difference between the bimodal and monomodal distributions is that the bimodal distribution "leaks" less radiation from the sides, the difference being absorbed by the large drops.

Decreasing the concentration of small droplets by an order of magnitude (C.5/10) to a value of 10 cm^{-3}, decreases the value of cloud reflectance to only 6% while increasing the value of absorptance to 9.6%. As with plane-parallel clouds, a decrease in small droplet concentration increases cloud absorptance. The fraction of incident energy exiting the sides has increased to 64.7% and the total radiation scattered out of the atmosphere has decreased to 14.6%.

Decreasing the concentration of small droplets another order of magnitude (C.5/100) to a value of 1 cm^{-3} further decreases the value of cloud reflectance to 3.1% and increases the value of absorptance to 10.1%. The decrease in cloud droplet concentration has also significantly increased the value of cloud transmittance. Only 48.7% of the incident energy is scattered out the cloud sides and only 8.1% is scattered out of the earth's atmosphere in this case. The value of cloud reflectance is only slightly greater than for the Rain 50 monomodal distribution, and the value of absorptance is also nearly equivalent to the large drops alone. Therefore, as in plane-parallel clouds, a small-droplet concentration of 1 cm^{-3} does not significantly alter the radiative characteristics of finite clouds.

Next, decreasing the horizontal dimensions of the cloud to 0.42 km square for the C.5/10 + Rain 50 distribution results in a decrease in the value of reflectance from 6 to 3.3%, while the transmittance decreases to 9.1%, and the absorptance changes from 9.6 to 7.9%; the energy scattered out of the earth's atmosphere decreases from 14.6 to 11.2% and the radiation scattered out of the cloud sides increases from 64.7 to 79.7%. Likewise, an increase in the width of the cloud to 1.68 km square increases the value of cloud reflectance to 11.7%, increases the value of transmittance to 31.6%, increases the value of absorptance to 11.1%, increases the value of energy scattered out of the atmosphere to 18.2%, and decreases the energy lost out of the cloud sides to 45.6%. This is a natural result of a decrease in the importance of the edge effects for the larger cloud.

For the 0.42 km square cloud, an increase in cloud thickness from 0.96 to 2.02 km leads to no change in the value of cloud reflectance. However, the value of transmittance is decreased from 9.1 to 0.2%, and the value of absorptance is increased from 7.9 to 9.5%, and the amount of radiation lost through the cloud sides changes from 79.7 to 87%. Doubling the horizontal cloud dimension to 0.84 km square increases the values of *R*, *T* and *A*. Increasing the cloud thickness to 3.21 km increases the value of absorptance to 14%. Therefore, for both monomodal and bimodal drop size distributions, increasing cloud thickness does not appreciably alter the value of cloud reflectance, but increases the value of absorptance and energy exiting the cloud sides. Increasing the horizontal size of the cloud increases both reflectance and absorptance, while decreasing the amount of energy "leaking" out the cloud sides.

5.4.4 Simulation of Cloud Radiative Characteristics Deduced from Aircraft Flux Observations

The Monte Carlo computing techniques developed for studying the radiative characteristics of finite clouds lend themselves to a slightly different, but highly complementary, application in this research. Namely, by simulating the irradiance characteristics of a finite cloud we can quantitatively assess the impacts of only partially sampling these patterns from an instrumented aircraft. In particular, aircraft observations above the cloud measure a value of incoming radiative flux proportional to $\cos\theta$. However, a cubic cloud intercepts radiation proportional to $\cos\theta + \sin\theta$, therefore, for $\theta > 0$ more energy is available for absorption than for plane-parallel clouds. On the other hand, significant radiation exits the sides of finite clouds which is not the case for semi-infinite clouds.

The simplest case in which observed irradiances may be converted into the R, T and A characteristics of the layer, is when the layer has a homogeneous irradiance field. The "infinite cloud" most often assumed in radiative transfer studies is an excellent example of this simplest case. One may then arbitrarily select the sampling area and use Eq. (5.5) to convert the irradiance data into the more fundamental radiative characteristics of the layer, i.e.,

$$R = \frac{F\uparrow(\tau = 0)}{F\downarrow(\tau = 0)}, \quad (5.5a)$$

$$T = \frac{F\downarrow(\tau = \tau_0)}{F\downarrow(\tau = 0)}, \quad (5.5b)$$

$$A = \frac{[F\downarrow - F\uparrow]_{(\tau=0)} - [F\downarrow - F\uparrow]_{(\tau=\tau_0)}}{F\downarrow_{(\tau=0)}}. \quad (5.5c)$$

See Fig. 5.1 for the relationship of the irradiance ($F\uparrow$ and $F\downarrow$) to the finite cloud geometry, and the symbol τ refers to optical depth. Poellot and Cox (1977) have quantified and reported the ramifications of only partially sampling a statistically homogeneous cloud field. They concluded that low-frequency sampling biases may represent a larger source of uncertainty than instrumental error for sample volumes less than

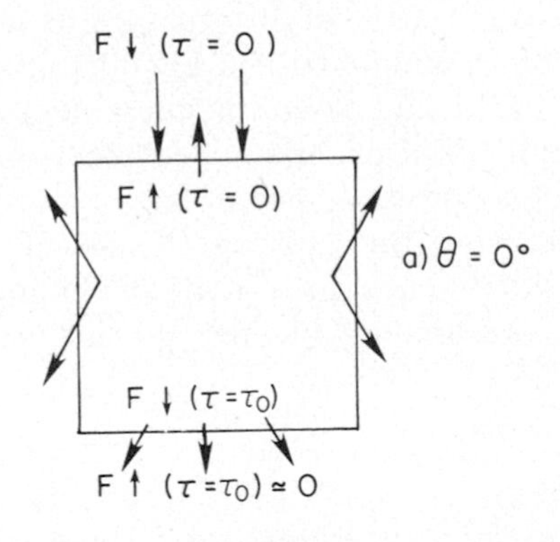

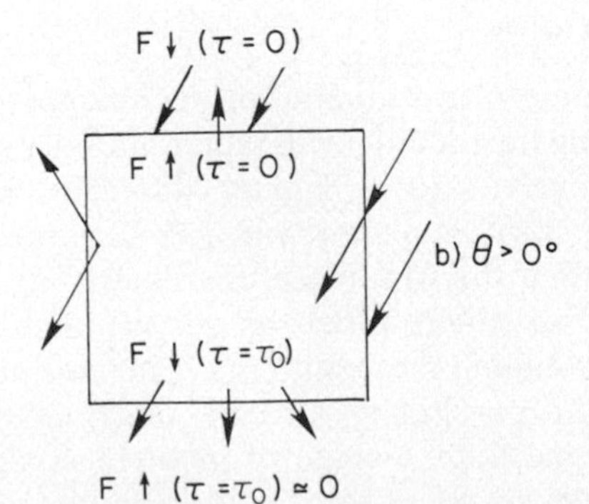

FIG. 5.1. A model finite cloud with upward ($F\uparrow$) and downward ($F\downarrow$) flux values at cloud top ($\tau = 0$) and cloud base ($\tau = \tau_0$). (a) solar zenith angle 0° and (b) solar zenith angle > 0°.

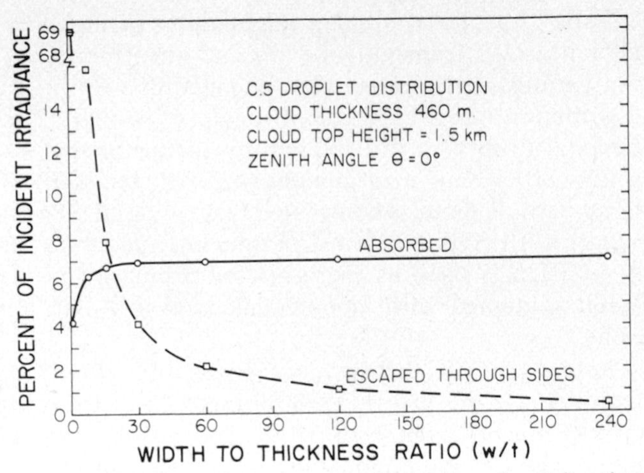

FIG. 5.2. Relationships between relative size of a finite cloud and absorption and energy escaping through the side faces of the cloud element.

100 km in length. This conclusion suggested that irradiance measurements from aircraft could be usefully employed to determine bulk radiation budgets of relatively large volumes.

The pitfalls of applying Eq. (5.5) to irradiance observations which meet neither the infinite cloud nor the statistical homogeneity criteria are well recognized by observers. However, nature seldom cooperates by providing "perfect" cloud layers for us to observe. Therefore, we shall now quantitatively explore differences between infinite cloud characteristics and those of finite clouds or edges of large semi-infinite cloud layers.

For the finite cloud, the results below are an extension of work reported by Davies (1978) for visible wavelengths. We have expanded the spectral interval to include all solar wavelengths, and compared the magnitudes of the energy lost through the sides of the finite cloud to the absorption.

In the expression for fractional absorptance given in Eq. (5.5c), it is implicitly assumed that the horizontal gradients of the horizontal fluxes are zero (i.e., $\partial F_x/\partial x = \partial F_y/\partial y = 0$). Although the finite cloud gives a vivid example of violating this assumption, there is a point when the width of a finite cloud element is sufficiently large that its bulk radiative properties must approach those of a semi-infinite cloud layer. The nonzero net horizontal flux divergence from the sides of the finite cloud may be used as a measure of a cloud's "finiteness". The net vertical flux convergence minus the horizontal flux divergence is then the actual absorption. Fig. 5.2 illustrates how fast a specific finite cloud case approaches the bulk radiation characteristics of its infinite counterpart. In this example a C.5 droplet distribution was used with vertical cloud thickness of 0.46 km and cloud top at 1.5 km. The two curves represent the fraction of the incident

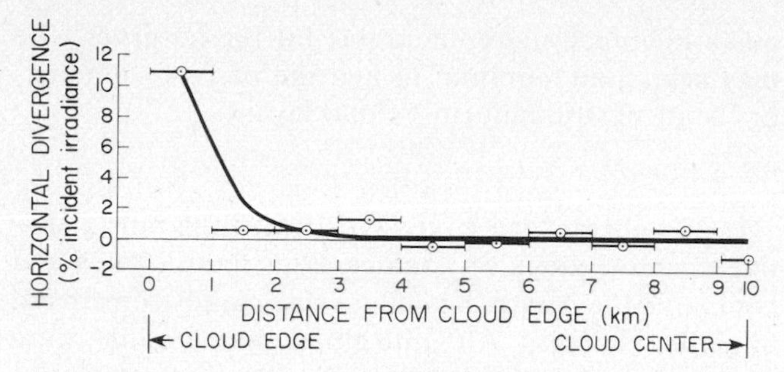

FIG. 5.3. Horizontal divergence of the horizontal components of total solar flux as a function of distance from cloud edge. The edge of the cloud is located at 0 km on the relative distance scale and the center of the cloud is at 10 km.

energy, for $\theta = 0°$, exiting the sides of the cloud and the actual fractional absorptance of the cloud layer. By the time the cloud attains a horizontal dimension of ~ 108 km ($w/t = 240$), less than 0.5% of the incident energy is escaping from the sides of the cloud. This value may then be compared with the actual absorption of 7%. Another way of expressing this comparison is to say that if Eq. (5.5c) for A were applied to the integrated irradiance values at cloud top and cloud base, the deduced value of A would be in error by an absolute amount of 0.5%. At the other extreme with $w/t = 8/1$ the energy escaping the sides would be 2.5 times the actual absorptance.

While Fig. 5.2 presents a comparison of the bulk finite cloud absorptance with the total amount of energy escaping through the cloud sides, it gives little indication of the variation in the divergence of the horizontal components of the flux within the cloud. A specific question now arises: "How far from the edge of a cloud toward its interior must one go before Eq. (5.5c) may be validly applied on a limited volume of the cloud?" Fig. 5.3 shows the magnitude of the divergence in the horizontal component of the flux as a function of distance (within the cloud) from the cloud edge. A cloud width of 20 km and the same cloud characteristics described above were used. The points plotted were generated from the Monte Carlo hybrid model (Davis *et al.*, 1979a,b) with a horizontal resolution of 1 km as indicated by the horizontal bars in the figure. The continuous line is the authors' impression of the probable fit to the plotted data points which possess considerable statistical noise. However, one can readily see that the horizontal component of the divergence clearly vanishes within 2–3 km of the cloud edge. This means that Eq. (5.5c) could be confidently applied to observations of the vertical fluxes for volume elements at least 2–3 km within the cloud. This figure further identifies the extent of the cloud affected by the side-escaping radiation depicted in Fig. 5.2. Essentially, most all the energy escaping from the sides of the cloud was incident within 2–3 km of a side face.

We now examine the spectral distribution of the energy escaping through the side faces of a finite cloud element and of the absorptance of that element. Table 5.6 shows the percentage of incident energy at cloud top which exits the cloud top, cloud base and cloud sides as a function of spectral region for the C.5 drop size distribution. The cloud is 3.36 km in both horizontal directions and 0.46 km thick ($t/w = 1/8$); cloud top is at 1.45 km. The cloud reflectance is largest for the shorter wavelengths (~50%) and smallest for the longer wavelengths (~0%), with an energy-weighted average reflectance of about 49%. Approximately 32% of the energy for $\lambda < 0.7$ μm exits cloud base; the total spectral average value of energy exiting cloud base is 29.2% of the incident irradiance. Nearly 17% of the energy exits the sides at the shorter wavelengths with negligible absorption occurring in these same spectral intervals. Only negligible amounts of energy exit the cloud sides for wavelengths $\gtrsim 2.8$ μm, with nearly the total incident energy at these wavelengths being absorbed. Considering the total solar spectral region, 15.8% of the incident energy exits the cloud sides while only 6.3% is absorbed. Note that the percentage of energy exiting the cloud sides for the total

TABLE 5.6. Percent of incident energy at cloud top which exits the cloud top, cloud base and cloud sides along with energy absorbed within the cloud as a function of wavelength region.

	Wavelength region (μm)									
	<0.7	0.76	0.95	1.15	1.4	1.8	2.8	3.3	6.3	Total
C.5 droplet distribution; $t/w = 1/8$; $\theta = 0°$										
Cloud top	50.8	50.9	51.7	49.1	36.5	26.7	0.4	0.4	0.6	48.8
Cloud base	32.3	32.4	28.3	25.8	15.6	8.7	0.0	0.0	0.0	29.2
Cloud sides	16.9	16.7	15.8	15.0	11.6	8.8	0.16	0.17	0.22	15.8
Absorptance in cloud	0.0	0.0	4.2	10.1	36.3	55.8	99.4	99.4	99.2	6.3
Best droplet distribution; $t/w = 1/2$; $\theta = 30°$										
Cloud top	26.2	26.1	29.5	28.6	26.1	24.3	4.4	1.3	1.4	26.6
Cloud base	22.7	22.6	19.2	18.2	16.0	15.0	3.6	1.7	1.4	21.0
Cloud sides	51.2	51.1	49.3	47.6	43.3	40.8	10.8	4.0	4.0	49.7
Absorptance in cloud	0.0	0.2	2.0	5.6	14.6	19.9	81.2	93.0	93.2	2.7

solar spectrum is only slightly smaller than the percentage exiting the cloud sides in the visible region.

The above results were for the case of a solar zenith angle of $\theta = 0°$. As the solar zenith angle increases, the sides of the cloud are also illuminated, with energy proportional to $\Delta z \sin \theta$. However, the energy incident through the cloud sides has a much shorter optical path than that incident through cloud top. Fig. 5.1 illustrates this situation. The downward flux at cloud base and the upward flux at cloud top are increased over those values which would result if all of the energy were incident at cloud top.

Table 5.6 shows the results of a second calculation for a cloud 0.84 km square and 0.46 km thick ($t/w \approx 1/2$) at a solar zenith angle of $\theta = 30°$ using the Best droplet distribution. At the smaller wavelengths about 51% of the incident energy upon both cloud top and cloud sides, (i.e., $\Delta x \cos \theta + \Delta z \sin \theta$) exits the cloud sides, while the average value (energy averaged over the entire solar spectrum) is about 50%. Once again, the percentage of the incident irradiance exiting the cloud sides at short wavelengths is an excellent approximation for the percentage of the spectrally averaged value escaping through the cloud sides.

Now let us see how we may apply this knowledge of the partitioning of the energy "leaking" out the sides of a finite cloud element. In the case of conservative scattering the energy "leaking" out the sides of the finite element is equal to the convergence of the vertical fluxes. Therefore, by computing or observing the convergence of the vertical fluxes for wavelengths < 0.7 μm, we simultaneously have an estimate of the percentage of incident light "leaking" out the sides of the cloud at this wavelength. This, of course, assumes that there is no actual absorption in the shorter wavelength interval.

Taking advantage of the fact that this percentage is only a slight overestimate of the percentage of the total energy "leaking" out the sides of the cloud, we may estimate the absorption of the finite element by first applying Eq. (5.5c) to a set of visible flux observations/calculations and determining the vertical flux convergence in this spectral region of nearly conservative scattering. This determines the "leakage" out of the sides of the cloud in a direct sense for the visible spectrum and implicitly for the total spectrum as noted above. We next apply Eq. (5.5c) to a set of vertical flux observations for the total solar spectrum and subtract out the total energy "leakage" as approximated from the visible bandpass convergence value.

The dual bandpass technique outlined above allows one to quantitatively assess the amount of energy exiting the side of a finite volume element of a cloud. It also enables one to make a first-order correction for this "leakage" on absorptance values determined from broadband irradiance observations. This should make life much more endurable for the observer who may search ad infinitum to find the perfect "infinite" or "statistically uniform" cloud layer.

5.5 Summary

In this chapter we have investigated the bulk radiative characteristics of homogeneous finite clouds and compared them with the infinite slab clouds considered in earlier chapters. All finite cloud computations were made using a two-stage Monte Carlo radiative transfer model described by Davis *et al.* (1979a,b).

For the range of cloud geometries reported in this chapter the three basic broadband cloud radiative characteristics, reflectance, transmittance and absorptance, were highly dependent upon the cloud geometry. Most of the variations were the result of different amounts of energy escaping from the sides of the clouds with differing geometries. In the absolute sense broadband cloud reflectance appeared most sensitive to geometric properties with transmittance and absorptance responding in a secondary, but nevertheless significant, manner.

For plane-parallel clouds the broadband absorptance was nearly invariant to the choice of drop size distribution; however, for finite clouds absorptance was noted to vary by as much as a factor of 2 depending upon the drop size distribution. Most of the sensitivity of cloud reflectance to the monomodal size distribution occurred for wavelengths $\lambda > 0.7$ μm suggesting a repartitioning of the energy between reflectance and absorptance at the longer solar wavelength. Finite cloud radiative characteristics reacted to the bimodal drop size distribution in a manner similar to their plane-parallel counterparts; the small-droplet size concentration dominated the reflectance and the addition of the large-droplet species increased absorptance.

The relative amounts of energy, and their spectral distributions, escaping from the various surfaces of a finite cloud element have been examined. It was shown that for a finite cloud element with a thickness to width ratio of 1/8 the energy escaping from the sides of the cloud was approximately 2.5 times the absorptance; a cloud with a thickness to width ratio of 1/240 had only 0.5% of the incident energy escaping through the sides. It was further shown that most of the energy escaping through the cloud sides was incident within 2–3 km of the side through which it escapes. The computations further revealed that the magnitude of the percentage of incident energy exiting the sides for wavelengths < 0.7 μm was only a slight overestimate of the magnitude of the percentage of energy integrated over all solar wavelengths which exited the cloud sides. This interesting coincidence of the magnitudes of the relative amounts of energy appears to be useful as an observational strategy for deducing the bulk radiative characteristics of heterogeneous cloud layers.

CHAPTER 6

The Effect of Vertical and Horizontal Cloud Microstructure Inhomogeneities upon the Radiative Characteristics of Cloud Layers

RONALD M. WELCH, STEPHEN K. COX AND JOHN M. DAVIS

The previous chapters have been primarily concerned with clouds of uniform cloud microstructure. However, in Chapter 2 vertical variations of liquid water content were analyzed. It was found that the cloud radiative characteristics were primarily a function of cloud optical depth rather than cloud vertical structure. Clouds with liquid water contents which decreased with increasing cloud depth gave essentially the same results as for clouds in which the liquid water content increased with increasing cloud depth; both cases gave essentially identical results to a homogeneous cloud with an average value of liquid water content. While the bulk cloud radiative properties were invariant to vertical cloud structure, local radiative characteristics within the cloud were highly sensitive to such variations. Local cloud heating rates were shown to vary substantially with variations in cloud microstructure.

6.1 Vertical cloud structure

The importance of side and edge effects in determining the bulk radiative properties of finite clouds suggest that the results noted in the opening paragraph for plane-parallel clouds may not be applicable to finite clouds. Table 6.1 shows values of cloud reflectance *(R)*, transmittance *(T)* and absorptance *(A)* as a function of solar zenith angle and vertical cloud structure for various finite cloud cases. (The precise definitions of *R*, *T* and *A* may be found on pp. 70 and 73.) The cloud is assumed to be 0.84 km square and 0.96 km thick with cloud top at 1.95 km. Two cases are considered using the C.5 drop size distribution. First we consider that the full liquid water content (drop concentration) as given by the C.5 distribution exists in the upper third of the cloud. The center third of the cloud is assumed to have 80% of this value, while in the lower third of the cloud a value of 60% is assumed. Therefore, in the lowest third of the cloud, the drop concentration is assumed to be 60 cm^{-3}, compared to 100 cm^{-3} in the upper portion of the cloud and 80 cm^{-3} in the center portion of the cloud.

For a solar zenith angle of $\theta = 0°$, the first liquid water content distribution gives a value of cloud reflectance of 30.7%, with 39% of the extraterrestrial energy available for cloud incidence scattered back out of the atmosphere. In terms of the energy incident at the top of the atmosphere 24.2% is reflected at cloud top; 2.1% of the incident energy at cloud top is transmitted through the cloud base and 5.7% is absorbed. Therefore, 61.5% of the energy exits the cloud sides. The maximum droplet concentration at cloud top yields nearly as large a value of cloud reflectance as the homogeneous cloud (Table 5.4 C.5 droplet distribution). The homogeneous cloud has a value of reflectance of 31.2% and "leaks" 61.4% of the energy out

TABLE 6.1. Percent cloud reflectance *(R)*, transmittance *(T)* and absorptance *(A)* as a function of solar zenith angle and vertical cloud structure as discussed in the text. For a detailed explanation of the normalization used in columns A, B and C see pp. 70 for $\theta = 0°$ and 73 for $\theta > 0°$.

Layer (N cm^{-3})	Zenith angle (deg)	Δx (km)	Δy (km)	Δz (km)	Cloud top (km)		A Top of cloud	B Top of atmosphere	C Remote sensing (DR)
100									
80	0	0.84	0.84	0.96	1.95	*R*	30.7	24.2	39.0
60						*T*	2.1	1.7	
						A	5.7	4.5	
60									
80	0	0.84	0.84	0.96	1.95	*R*	21.6	17.1	34.4
100						*T*	2.5	2.0	
						A	5.3	4.2	
100									
80	60	0.84	0.84	0.96	1.95	*R*	20.1	15.1	31.4
60						*T*	15.6	11.8	
						A	3.2	2.4	
60									
80	60	0.84	0.84	0.96	1.95	*R*	17.0	12.8	31.1
100						*T*	13.2	10.0	
						A	3.2	2.4	

the cloud sides. The homogeneous cloud, however, has a slightly larger value of cloud absorptance, 5.8%.

The second case considers a cloud microstructure in which the droplet concentration at cloud top is assumed to be 60 cm^{-3} while that at cloud base is now 100 cm^{-3}. For this case the value of cloud reflectance decreases to 21.6% with 34.4% of the extraterrestrial incident energy transmitted out of the atmosphere. The values of cloud transmittance and absorptance are quite similar. This cloud leaks 70.6% of the incident energy out the cloud sides. Clearly, vertical structure, in contrast to the plane-parallel case, significantly affects the bulk radiative characteristics of finite clouds. These effects for small clouds are primarily associated with the value of cloud reflectance. The cloud in which the droplet concentration increases from cloud top to cloud base permits a greater percentage of the radiation (about 10%) to exit the cloud sides, at the expense of reflection through the cloud top. As the thickness-to-width ratio is decreased (as the cloud approaches the plane-parallel limit), the differences between these two cases decrease. A decrease in the energy exiting the cloud sides in the finite cloud case contributes primarily to increased values of cloud reflectance as the cloud approaches the plane-parallel limit. However, the values of cloud transmittance and absorptance also increase in the plane-parallel limit (see also Fig. 5.1).

The next set of calculations is for a solar zenith angle of $\theta = 60°$ in which the majority of the solar energy is incident upon the cloud sides. The cloud for which the droplet concentration is maximum (100 cm^{-3}) at cloud top once again has the largest value of cloud reflectance (20.1% compared to 17% for the opposite case). However, the value of cloud absorptance is the same for both cloud cases (3.2%), and the values of radiation reflected out of the earth's atmosphere are nearly equal. The cloud with the smallest drop concentration at cloud base has the largest value of cloud transmittance. Although not shown here, the basic conclusions discussed above were verified for other drop size distributions, cloud geometries, cloud top heights and microstructure variations.

6.2 *Holes in finite clouds*

Platt (1976) has shown that at a given level within the cloud body, while the drop size distribution remains unchanged, there are various inhomogeneities in cloud droplet concentration (and, therefore, in optical density and liquid water content). Jonas and Mason (1974) reported that large "holes" may develop within the cloud body. The purpose of the present section is to investigate how the presence of such holes influences the radiative characteristics of finite clouds.

Van Blerkom (1971) investigated the effect of cloud-top structure, in the form of horizontal striations, on the diffuse reflection of radiation. Isotropic scattering of the cloud droplets was assumed. Significant differences between these clouds with periodic inhomogeneities in one horizontal direction and the plane-parallel case were reported. Photons reflected in directions nearly normal to the cloud were, of course, affected less by the presence of striations than those scattered in more oblique directions.

Wendling (1977) also considered scattering of photons from horizontal periodic striations at a wavelength of $\lambda = 0.55$ μm. Cloud albedo and reflected radiance were calculated as functions of cloud drop size distribution, optical thickness, solar geometry and type of cloud striations. Cloud albedo is reported lower for striated clouds than for a plane-parallel cloud of the same mean optical thickness, with differences of about 20% for deep striations at a solar zenith angle of $\theta = 0°$. The choice of drop size distribution was shown to be as important a consideration upon the cloud albedo as variations in the type of striations. Variations in the width of the striations had a much smaller effect on the value of cloud albedo than variations in cloud surface structure.

Table 6.2 provides calculations of cloud reflectance, transmittance and absorptance for various finite cloud models with holes, i.e., regions within the cloud body with no droplets present. The C.5 droplet size distribution is assumed for the cloud body. Similar results are obtained for other monomodal and bimodal drop size distributions. Due to previous intercomparison studies of the effects of drop size distribution upon the cloud radiative characteristics, results only for the C.5 size distribution are presented as a representative case. The hole in the cloud is assumed to be square in shape and to occupy the center third of the cloud body. For the first three cases studied the hole is assumed to extend from cloud top to cloud base. The first cloud case given in Table 6.2 is for a cloud 0.84 km square and 0.96 km thick with a cloud top at 1.95 km. Solar zenith angle is $\theta = 0°$ in all cases. As reported by Wendling, the effect of cloud striations upon the cloud radiative characteristics is greatest at large solar zenith. For this cloud 0.84 km square, the hole is 0.28 km square located at cloud center.

Table 5.4 showed that a homogeneous cloud of the same dimensions had a value of cloud reflectance of 31.2% with 40.1% of the incident extraterrestrial energy scattered back out of the earth's atmosphere. The value of cloud transmittance was 1.6% with 5.8% of the incident energy (at cloud top) absorbed and 61.4% scattered out the cloud sides. The presence of a hole at cloud center decreases the value of cloud reflectance to 25.6% and decreases the radiation scat-

TABLE 6.2. Percent cloud reflectance (R), transmittance (T) and absorptance (A) as a function of cloud geometry for finite clouds with "holes", as discussed in the text. For a detailed explanation of the normalization used in columns A, B and C see p. 70.

Type	Zenith angle (deg)	Δx (km)	Δy (km)	Δz (km)	Cloud top (km)		A Top of cloud	B Top of atmosphere	C Remote sensing (DR)
Hole in center 3 sections	0	0.84	0.84	0.96	1.95	R	25.6	20.2	33.7
						T	13.5	10.7	
						A	5.2	4.1	
Hole in center 3 sections	0	0.42	0.42	0.96	1.95	R	14.9	11.8	26.9
						T	11.6	9.1	
						A	3.8	3.0	
Hole in center 3 sections	0	1.68	1.68	0.96	1.95	R	37.9	29.9	39.5
						T	17.2	13.6	
						A	6.7	5.3	
Pocket in top section	0	0.84	0.84	0.96	1.95	R	27.4	21.6	38.2
						T	2.8	2.2	
						A	5.8	4.6	
Pocket in top section	0	1.68	1.68	0.96	1.95	R	42.8	33.8	44.6
						T	8.0	6.3	
						A	7.8	6.2	
Pocket in center section	0	0.84	0.84	0.96	1.95	R	30.7	24.2	39.5
						T	2.2	1.7	
						A	5.8	4.5	
Pocket in lower section	0	0.84	0.84	0.96	1.95	R	31.3	24.7	39.9
						T	1.9	1.5	
						A	5.8	4.6	

tered back out of the atmosphere to 33.7%. The value of cloud transmittance increases to 13.5%. The hole occupies about 11% of the cloud area, and the increase in cloud transmittance is primarily due to radiation which propagates directly down through the hole without scattering; however, comparison of the transmittance values indicates that transmittance is also increased by photons which scatter into the downward direction through the sides of the hole. The value of cloud absorptance has decreased to 5.2%, and the amount of radiation leaking out the cloud sides has decreased to 55.7%.

Two other cloud sizes are considered, one with cloud width halved and the other with cloud width doubled. In each case the hole occupies the center section of the cloud and is 1/9th of the cloud top area. The cloud 0.42 km square has a value of cloud reflectance of 14.9% with 26.9% of the incident extraterrestrial radiation scattered out of the atmosphere. The larger 1.68 km square cloud has a cloud reflectance of 37.9% with 39.5% of the radiation scattered out of the atmosphere. Corresponding values of cloud reflectance for homogeneous clouds are 18.5 and 44.8%, respectively, with 31.7 and 46%, respectively, scattered out of the earth's atmosphere. The value of cloud transmittance for the "tall" cloud ($t/w \approx 2/1$) is 11.6%, while that for the "flattened" cloud ($t/w \approx 1/2$) is 17.2%. Most of the radiation exiting the cloud base for the tall cloud is energy which is propagating straight down through the hole without interaction. Photons interacting with droplets within the cloud body are far more likely to be scattered out of the cloud sides than out of the cloud base, even with the presence of the hole. The value of cloud absorptance increases with increasing cloud width (i.e., with decreasing value of the t/w ratio), for a cloud of constant thickness. Table 6.2 also shows results for clouds with "pockets" of clear air occupying the center portion of the cloud, which are assumed to exist in the upper third, center third, or lower third of the cloud. The horizontal area of the pocket is 11% of the cloud top area. The pocket is first located in the top third of the cloud. As would be expected the cloud with the pocket in the top section has a larger value of cloud reflectance (27.4%) than does the cloud with the hole all the way through the center section (25.6%), but a value lower than that of the homogeneous cloud (31.2%). A similar relationship exists among the values of incident extraterrestrial radiation scattered out of the atmosphere. The value of cloud absorptance is equivalent to the value found in the homogeneous cloud but greater than the absorptance in the cloud with the clear center column. Slightly more radiation is scattered through the cloud sides than in the homogeneous cloud case.

Similar behavior is shown for the 1.68 km square cloud. The pocket is 1/9th of the upper surface area and extends to a cloud depth of 0.32 km. The cloud with the pocket at the top has a greater value of reflectance and absorptance than does the cloud with the hole through its center. It also scatters more radiation out of its sides than does the homogeneous cloud. These features appear to be independent of the actual cloud width.

The final two calculations, carried out for clouds with 0.84 km square tops, assume that the pocket is displaced first to the center of the cloud, and then to the cloud base layer. As the pocket is lowered in position within the cloud, the value of cloud reflectance and the value of radiation scattered out of the atmosphere are increased. The value of cloud absorptance remains relatively unchanged. However, the transmittance decreases as the position of the pocket within the cloud is lowered. While the cloud with the pocket at the top scatters slightly more radiation out of the cloud sides than does a homogeneous cloud, clouds with the pocket in the center or lower sections scatter slightly less radiation out of the cloud sides than does a homogeneous cloud.

6.3 Holes in infinite clouds

The previous section has considered the effects of holes, or regions within the cloud without droplets, upon the finite cloud radiative characteristics. The present section extends this treatment to the semi-infinite cloud. In these models it is assumed that holes, extending through the cloud or located at some position within the cloud body, are repeated at regular intervals.

Table 6.3 shows values of reflectance, transmittance and absorptance for various types of horizontally infinite clouds with regularly dispersed holes or clear regions. The Best droplet distribution for a 0.1 g m^{-3} liquid water content was used in all cases. Cases 1–3 represent 0.46 km thick "infinite" clouds with regularly spaced vertical clear columns, which comprise 11% of the total cloud. Results for zenith angles of 0°, 30° and 60° are shown. For the 0° zenith angle case, 48.4% of the incident cloud top energy is reflected back out the top of the cloud and 35.8% of the extraterrestrial incident energy is scattered back out of the top of the atmosphere. This compares with 55.1% reflectance and 40.7% scattering out of the atmosphere by the homogeneous cloud. Transmittance is increased for the porous cloud to 47.3% as compared to 40.1% for the homogeneous cloud, and absorptance is slightly less when compared to the homogeneous cloud (4.3% compared to 4.8%). As the zenith angle is increased to 30° and to 60°, the same general comparisons may be made and similar conclusions may be drawn regarding transmittance and reflectance results. For a 60° zenith angle, the array of cloud/clear area reflects 57.5% and transmits 36.6% of the incident cloud energy,

TABLE 6.3. Percent cloud reflectance (*R*), transmittance (*T*) and absorptance (*A*) for horizontally infinite clouds with various "holes" or "clear regions". For a detailed explanation of the normalization used in columns A, B and C see p. 70 for $\theta = 0°$ and 73 for $\theta > 0°$.

Case	Type	Zenith angle (deg)	Δz (km)	Cloud top (km)		A Top of cloud	B Top of atmosphere	C Remote sensing (DR)
1	Holes in vertical columns	0	0.46	1.45	*R* *T* *A*	48.4 47.3 4.3	37.6 36.8 3.3	35.8
2	Hole in center vertical column	30	0.46	1.45	*R* *T* *A*	54.8 40.9 4.3	42.2 31.5 3.1	40.2
3	Hole in center vertical column	60	0.46	1.45	*R* *T* *A*	57.5 36.6 5.9	40.9 26.0 4.2	38.3
4	Hole in center vertical column	0	0.96	1.95	*R* *T* *A*	57.0 37.4 5.6	44.9 29.5 4.4	42.9
5	Hole in center vertical column	30	0.96	1.95	*R* *T* *A*	69.0 22.8 8.2	53.9 17.8 6.4	51.8
6	Hole in center vertical column	60	0.96	1.95	*R* *T* *A*	76.1 16.1 7.8	51.8 10.9 5.3	49.0
7	Pockets in top third of cloud	0	0.46	1.45	*R* *T* *A*	53.7 41.4 4.9	41.8 32.2 3.8	39.7
8	Pockets in middle third of cloud	0	0.46	1.45	*R* *T* *A*	53.8 41.3 4.9	41.9 32.1 3.8	39.8
9	Pockets in bottom third of cloud	0	0.46	1.45	*R* *T* *A*	53.5 41.7 4.9	41.6 32.1 3.8	39.5

while the corresponding homogeneous cloud values are 69.6% and 27.5% respectively. However, in this case, the porous cloud absorbs a relatively greater fraction of the incident cloud energy, 5.9% as compared to 2.9% for the homogeneous cloud. This greater fractional absorptance may be attributed to the ability of radiation incident on a hole at the level of the cloud top to propagate in a straight line path (specified by the larger zenith angle) and enter the cloud at a point well below cloud top; as a consequence the radiation scattered out the cloud top is less than for the homogeneous case. The result is an enhancement of the fractional absorptance value. For the thicker array (Δz = 0.96 km), with $\theta = 0°$, the reflectance has been decreased compared to an equally thick homogeneous cloud, from 69.5 to 57.0%. Transmittance has correspondingly increased from 20.4 to 37.4%, and the porous cloud absorbs 5.6% of the incident radiation compared to 10% in the "solid" cloud. As the zenith angle is increased to 30° and to 60°, the expected increases in reflectance values and decreases in transmittance values are observed for the porous cloud. With a 60° zenith angle the thicker porous cloud shows values of R, T and A of 76.1%, 16.1% and 7.8% with 49.0% scattered back out of the atmosphere. These compare to R, T and A values in the homogeneous counterpart of 80.4%, 15.9% and 3.7% and 63.4% scattered back out of the top of the atmosphere. Clearly the presence of the "holes" has a significant effect on the bulk cloud absorptance by allowing a relatively deeper penetration of incident radiation into the cloud at larger zenith angles.

Cases 7, 8 and 9 in Table 6.3 consider pockets of clear air which do not penetrate the entire thickness of the cloud. The progression for these 0° zenith angle cases is for the clear pocket to be initially in the top third of the infinite lattice, and subsequently to move down to the middle and bottom thirds of the array. The horizontal extent of the clear air pockets is again 11% of the total cloud area. In Section 6.2 we showed that for the finite cloud the R, T, A values exhibit a dependence on the position of the pocket; in contrast, there is no such dependence shown in the infinite cloud case. However, the reflectance is slightly less than in the homogeneous cloud case, 53.7% compared to 55.1%, while transmittance and absorptance values are slightly greater, 41.4% compared to 40.1% and 4.9% compared to 4.8%.

6.4 Random horizontal and vertical inhomogeneities in finite clouds

The present section simulates the situation reported by Platt (1976), with horizontal variations in liquid water content and droplet concentration. The drop size distribution at a given cloud height, however, was found to be invariant with horizontal position. For simplicity in the following calculations, a uniform droplet size distribution was assumed throughout the cloud. The cloud was divided into three vertical levels and 81 horizontal levels, or 243 "boxes" of equal size. A random number generator assigned the droplet number density (i.e., optical thickness and liquid water content) to each "box". The droplet single-scattering phase function and single-scattering albedo were assumed to be invariant with position within the cloud. The C.5 droplet size distribution was used and droplet number densities (concentrations) varied from about 0 to 100 cm^{-3}.

Table 6.4 shows the bulk radiative properties of clouds with random inhomogeneities for various solar zenith angles and cloud widths. Cloud thickness was 0.96 km in all cases and cloud top height was held constant at 1.95 km. The results indicate a gradual change in the bulk radiative parameters of the randomized clouds as a function of zenith angle. Cases 1–3 indicate that transmittance increases monotonically and absorptance decreases monotonically with zenith angle while reflectance decreases at a 30° zenith but then increases at a 60° solar zenith angle. The results of Table 6.4 also show the expected increase in the bulk radiative parameters with decreasing height-to-width ratios. This behavior was noted in the homogeneous finite cloud cases (see Tables 5.1–5.3).

The bulk radiative properties of the randomized clouds were found to be relatively invariant to the actual values assumed for the cloud droplet concentrations. For example, the calculation represented by Case 1 was repeated twice. Each subsequent run was identical except for a change in the random sequence which assigned the variable extinction coefficients. The resulting values of cloud reflectance were 17.2 and 18.0%, values of transmittance were calculated to be 7.6 and 7.2%, and absorptance values were found to be equal to 5.1% of the total cloud incident energy. Case 3 was similarly repeated (only once in this case), and values of R, T and A were 15.8, 20.3 and 2.8%, respectively. Comparison of these results with those in Table 6.4 shows no great deviations in the respective R, T and A values due to different distributions of liquid water content throughout the clouds.

A comparison was also made between the nonhomogeneous clouds represented in Table 6.4 and their homogeneous counterparts, in which uniform values of liquid water content and volume extinction coefficients, equal to the average of the values used in the nonhomogeneous cases, were assigned throughout the clouds. These results indicate some slight differences between bulk radiative properties of the nonhomogeneous clouds and the corresponding homogeneous

TABLE 6.4. Percent cloud reflectance (R), transmittance (T) and absorptance (A) as a function of cloud geometry for random horizontal inhomogeneities in finite clouds. For a detailed explanation of the normalization used in columns A, B and C see pp. 70 for $\theta = 0°$ and 73 for $\theta > 0°$.

Case	Zenith angle (deg)	Δx (km)	Δy (km)	Δz (km)	Cloud top (km)		A Top of cloud	B Top of atmosphere	C Remote sensing (DR)
						R	18.3	14.5	29.7
1	0	0.90	0.90	0.96	1.95	T	7.6	6.0	
						A	5.2	4.1	
						R	14.7	11.5	25.8
2	30	0.90	0.90	0.96	1.95	T	15.4	12.1	
						A	4.0	3.1	
						R	16.3	12.3	27.9
3	60	0.90	0.90	0.96	1.95	T	18.4	13.8	
						A	2.8	2.1	
						R	30.3	23.9	35.3
4	0	1.80	1.80	0.96	1.95	T	16.3	12.9	
						A	6.6	5.2	
						R	37.0	29.2	37.3
5	0	2.70	2.70	0.96	1.95	T	21.6	17.1	
						A	7.3	5.8	
						R	37.4	28.1	36.9
6	60	2.70	2.70	0.96	1.95	T	21.5	16.2	
						A	4.1	3.1	

cases, primarily in the 0° zenith angle cases. Values of R, T and A for a homogeneous cloud with a liquid water content and a volume extinction coefficient equal to the average of those used in Case 1 are 19.3, 5.3 and 5.1%. The energy scattered back out of the top of the atmosphere was 31.3% of the extraterrestrial incident energy. Comparison with Case 1 of Table 6.4 shows the largest discrepancy is found in the values of transmittance. A similar comparison was made for the cloud of Case 4, for which R, T and A in the homogeneous cloud were found to be 31.1, 13.6 and 6.6% with 36% of the energy scattered out of the atmosphere; and for Case 5 which resulted in R, T and A values of 38.0, 18.8 and 7.3%, with 38.4% of the extraterrestrial energy scattered out of the top of the atmosphere. Again the largest differences between the bulk radiation properties of nonhomogeneous and "equivalent" homogeneous clouds are observed in the values of transmittance, with those of the nonhomogeneous clouds having slightly larger values. This type of discrepancy is diminished in the cases with zenith angles of 30° and 60°. For example, a calculation "equivalent" to Case 2 yields values of R, T and A of 15.0, 14.4 and 4.0%, with 26.4% of the incident extraterrestrial energy scattered out of the top of the atmosphere. A calculation "equivalent" to that of Case 6 results in values of R, T and A of 38.1, 21.0 and 4.1%, with 37.4% of the energy scattered out through the top of the atmosphere. These results are in good agreement with those of Case 6 in Table 6.4. The probable explanation for the relatively larger differences between the values of transmitted radiation in the 0° zenith angle cases may be the random creation of "holes" or "channels" which allows energy to pass through the cloud with few interactions. As the zenith angle increases, it becomes less likely that such holes might exist along the relatively longer diagonal path which non-interacting photons would take. The resulting values of transmittance for larger solar zenith angles would thus be much closer to those of the "equivalent" homogeneous cloud.

6.5 Random horizontal inhomogeneities in plane-parallel clouds

This section summarizes an analogous study for clouds with similar horizontal inhomogeneities but which are infinite in horizontal extent. Table 6.5 shows values (Cases 1, 2 and 3) of the bulk radiative parameters for a 0.46 km thick infinite cloud with randomly assigned extinction coefficients, for solar zenith angles of 0°, 30° and 60°. The method used to assign the volume extinction coefficients is analogous to that described in Section 6.4. Also shown in Table 6.5 (Cases 4, 5 and 6) are values of R, T and A for the "equivalent" horizontally homogeneous clouds, whose volume extinction coefficients have been set equal to the average value of the randomly distributed models. Inspection of Table 6.5 indicates only slight differences exist between the bulk radiative properties of the randomized "patchy" clouds and the horizontally homogeneous clouds with "equivalent" mean volume extinction coefficients. These differences are greatest at a zenith angle of 0° and are evident as enhanced

TABLE 6.5. Percent cloud reflectance (*R*), transmittance (*T*) and absorptance (*A*) for horizontally infinite clouds with random distributions of liquid water content and volume extinction coefficients, and for "equivalent" homogeneous clouds. For a detailed explanation of the normalization used in columns A, B and C see pp. 70 for $\theta = 0°$ and 73 for $\theta > 0°$.

Case	Type	Zenith angle (deg)	Δz (km)	Cloud top (km)		A Top of cloud	B Top of atmosphere	C Remote sensing (DR)
1	Random	0	0.46	1.45	*R*	34.1	26.5	25.1
					T	62.0	48.2	
					A	3.9	3.3	
2	Random	30	0.46	1.45	*R*	40.7	31.3	29.5
					T	55.7	42.9	
					A	3.6	2.8	
3	Random	60	0.46	1.45	*R*	55.7	41.3	39.4
					T	41.7	30.9	
					A	2.6	1.9	
4	Homogeneous	0	0.46	1.45	*R*	36.0	28.0	26.5
					T	60.1	47.8	
					A	3.9	3.3	
5	Homogeneous	30	0.46	1.45	*R*	41.1	31.7	30.1
					T	55.2	42.5	
					A	3.7	2.8	
6	Homogeneous	60	0.46	1.45	*R*	56.0	41.6	39.8
					T	41.4	30.7	
					A	2.6	1.9	

transmission with correspondingly reduced reflection in the randomized model as compared with the "equivalent" homogeneous cloud. At larger zenith angles these slight differences are diminished. This behavior is similar to that observed in randomized finite clouds which were discussed in the previous section. The reader is referred to the end of Section 6.4 for the probable explanation of such behavior.

6.6 *Finite cloud with cores (rainshowers) of large drops*

The previous calculations have assumed that the drop size distribution is uniform horizontally within the cloud. However, in highly convective clouds there may be regions (cores) in which large drops form. For simplicity it is assumed in this section that only large drops are present in the core regions and only small droplets are present in the surrounding regions. The C.5 small-particle distribution is assumed for the surrounding regions and the Rain 50 large-particle distribution for the core region.

The cloud body is partitioned as in Section 6.2, in which it was assumed that holes existed in the clouds. It is assumed that the core region covers 1/9th of the cloud top surface area and extends from cloud top to cloud base. Table 6.6 shows the results from finite clouds of various geometries with centered rainshower regions.

The first cases consider a cloud 0.84 km square, 0.46 km thick with cloud top at 1.45 km. Solar zenith angles are $\theta = 0°$, 30° and 60°. The values of cloud reflectance, transmittance and radiation scattered out of the atmosphere are relatively invariant to variations in solar zenith angle. However, the value of cloud absorptance decreases with increasing solar zenith angle when referenced to total cloud incident energy. This behavior is found for both finite and plane-parallel clouds and for clouds with variations of horizontal and vertical structure.

The next case increases cloud thickness to 0.96 km at $\theta = 0°$. The value of cloud reflectance remains nearly constant as the cloud thickness is increased, assuming that cloud width is kept constant. However, the total radiation scattered out of the atmosphere increases with increasing cloud thickness. These results are similar to those obtained for homogeneous finite clouds.

Comparing the values calculated for the cloud of similar geometry with the hole in the center section (Table 6.2) shows that the presence of a rain core instead of a hole increases the value of cloud reflectance from 25.6 to 26.6% and the radiation scattered out of the atmosphere from 33.7 to 36.1%. The value of cloud absorptance is increased from 5.2 to 6.4% and the value of cloud transmittance decreased from 13.5 to 7.2%. The presence of the rain core increases the amount of incident radiation scattered out of the cloud sides from 55.7 to 59.8%. These differences will be increased for clouds in which the core (either rain or clear) occupies a larger portion of the cloud body. It is not possible to include all such cases within the present work without making this monograph excessively lengthy. Therefore, the present results should only be used to indicate that the

TABLE 6.6. Percent cloud reflectance (R), transmittance (T) and absorptance (A) as a function of cloud geometry for finite clouds with cores (rainshowers) of large drops, as discussed in the text. For a detailed explanation of the normalization used in columns A, B and C see pp. 70 for $\theta = 0°$ and 73 for $\theta > 0°$.

Type	Zenith angle (deg)	Δx (km)	Δy (km)	Δz (km)	Cloud top (km)		A Top of cloud	B Top of atmosphere	C Remote sensing (DR)
Rain in center column	0	0.84	0.84	0.46	1.45	R	25.8	20.0	30.9
						T	21.7	16.9	
						A	5.0	3.9	
Rain in center column	30	0.84	0.84	0.46	1.45	R	23.4	18.0	29.0
						T	22.2	17.1	
						A	4.2	3.3	
Rain in center column	60	0.84	0.84	0.46	1.45	R	27.1	20.1	32.2
						T	21.8	16.1	
						A	2.9	2.2	
Rain in center column	0	0.84	0.84	0.96	1.95	R	26.6	21.0	36.1
						T	7.2	5.7	
						A	6.4	5.1	
Rain in center column	60	0.84	0.84	0.96	1.95	R	19.5	14.7	32.8
						T	13.2	9.9	
						A	3.7	2.8	
Rain in center column	0	1.68	1.68	0.96	1.95	R	39.3	31.0	41.6
						T	12.4	9.8	
						A	7.9	6.3	
Rain in center column	60	1.68	1.68	0.96	1.95	R	33.1	25.0	38.0
						T	14.9	11.3	
						A	4.7	3.6	
Rain in center column	0	3.36	3.36	0.96	1.95	R	49.2	38.8	44.4
						T	17.5	13.8	
						A	9.9	7.8	
Rain in top center region	0	1.68	1.68	0.96	1.95	R	42.4	33.4	44.4
						T	7.4	5.9	
						A	8.3	6.6	
Rain in top center region	60	1.68	1.68	0.96	1.95	R	34.3	25.8	38.8
						T	13.7	10.3	
						A	4.8	3.6	

presence of a rain core may increase the value of cloud reflectance and absorptance over that of a cloud with a hole in its center. However, the cloud with the rain core has a smaller value of cloud reflectance than does a homogeneous cloud.

For a solar zenith angle of $\theta = 60°$ the 0.84 km square cloud with thickness of 0.96 km has a smaller value of cloud reflectance and absorptance than does the cloud with $\theta = 0°$. This is in contrast to the cloud with a thickness of 0.46 km for which the value of cloud reflectance was nearly constant with variations in solar zenith angle.

Increasing the cloud width to 1.68 km square and to 3.36 km square increases the value of cloud reflectance, transmittance and absorptance as well as decreasing the amount of radiation leaked out of the cloud sides. The cloud with the rain core consistently has a larger value of cloud reflectance and absorptance than does the cloud with the hole in the center; but it consistently has a lower value of cloud reflectance than does the homogeneous cloud. These same conclusions are valid at other solar zenith angles.

As a final case the large drops are assumed to exist only in the upper third of the cloud core region, with the C.5 droplet distribution in the region below the large drops. A cloud 1.68 km square is assumed with cloud thickness of 0.96 km. For a solar zenith angle of $\theta = 0°$ the values of R, T and A are not greatly different from those obtained for a pocket (Table 6.2, Case 5) in this position. Therefore, for all practical considerations the presence of a pocket of clear air or a large-drop region, near cloud top, is indistinguishable.

6.7 Plane-parallel clouds with regions of rainshowers

In this section the effect of regularly repeated simulated rainshower columns on the bulk radiative properties of horizontally infinite clouds is examined. The C.5 droplet distribution is used for the basic infinite cloud which surrounds cores composed of monomodal and bimodal water droplet distributions. A solar zenith angle of 0° is maintained throughout all calculations. Table 6.7 shows the results of the

TABLE 6.7. Percent cloud reflectance (R), transmittance (T) and absorptance (A) for C.5 infinite clouds with regularly arrayed columns of rainshowers composed of various monomodal and bimodal droplet distributions. $\theta = 0°$ in all cases. For a detailed explanation of the normalization used in columns A, B and C see p. 70.

Type of column droplet distribution	Δz (km)	Cloud top (km)		A Top of cloud	B Top of atmosphere	C Remote sensing (DR)
Rain 50	0.46	1.45	R	50.8	39.1	37.3
			T	42.5	32.7	
			A	6.7	5.2	
Rain 50	0.96	1.95	R	63.3	50.0	48.1
			T	25.8	20.3	
			A	10.9	8.6	
Rain 10	0.46	1.45	R	50.0	38.4	36.6
			T	43.9	33.7	
			A	6.1	4.7	
C.5 + Rain 50	0.96	1.95	R	70.2	55.4	53.4
			T	18.4	14.5	
			A	11.4	9.0	
C.5/10 + Rain 50	0.96	1.95	R	65.6	51.8	49.9
			T	23.4	18.4	
			A	11.0	8.6	
C.5/100 + Rain 50	0.96	1.95	R	64.2	50.7	48.8
			T	24.7	19.5	
			A	11.0	8.8	

various calculations. In all cases the regularly arrayed rain core regions comprise 11% of the total cloud area. For a 0.46 km thick C.5 infinite cloud with regularly arrayed columns characterized by the Rain 50 droplet distribution, the reflectance is calculated at 50.8%, transmittance 42.5% and absorptance 6.7%. The energy scattered back out the top of the atmosphere is 37.3% of the extraterrestrial energy. Increasing the thickness of this cloud to 0.96 km results in 1) the expected increase in reflectance from 50.8 to 63.3%; 2) a decrease in transmittance from 42.5 to 25.8%; and 3) an increase in absorptance from 6.7% to 10.9%. The same thickness (0.96 km) of homogeneous C.5 clouds results in R, T and A values of 70.1, 19.1 and 10.8%, respectively. Thus, the presence of the Rain 50 cores has substantially reduced reflectance and increased transmittance but has not affected the value of bulk cloud absorptance. Replacing the Rain 50 cores with Rain 10 cores in the 0.46 km thick cloud has little effect on the bulk radiative properties of the cloud. Returning to the thicker cloud, and replacing the Rain 50 cores with a bimodal C.5 + Rain 50 distribution, has the effect of increasing reflectance to 70.2%, a value just slightly greater than the 70.1% found for the homogeneous C.5 cloud. Transmittance is slightly less and absorptance is slightly greater in the "arrayed" cloud than in the homogeneous cloud. If the concentration of the small droplets is decreased by an order of magnitude so that the core is composed of a C.5/10 + Rain 50 droplet distribution and reduced again by an order of magnitude to a C.5/100 + Rain 50 droplet distribution, the bulk radiative properties rapidly approach those of the cloud with the large-droplet core. This behavior is consistent with the bimodal discussion for relatively thin clouds (see Section 3.2.3).

6.8 Summary

This chapter has examined the effects of vertical and horizontal cloud microstructure inhomogeneities on the bulk radiative characteristics of clouds. It was shown that vertical variations in mean layer droplet concentrations have a significant effect on the values of reflectance in finite clouds. Finite clouds in which the droplet concentration increases from cloud bottom to cloud top have significantly higher reflectance values than finite clouds in which the droplet concentration increases from top to bottom. This effect appears to be a property of finite clouds since a similar variation in the vertical distribution of mean layer droplet distribution has little effect on the bulk radiative properties of infinite clouds (see Section 2.4.4).

The bulk radiative characteristics of finite clouds with columns or pockets of clear air were examined. As expected, reflectance and absorptance decrease and transmittance increases when a center column of the cloud is replaced by clear air. When pockets of clear air, whose vertical extent equaled one-third of the vertical cloud dimension, were placed at the top, middle or bottom of the center column of the cloud, a relatively small change was noted in the bulk radiative characteristics.

Similar results were found for infinite clouds with regularly arrayed columns or pockets of clear air at a solar zenith angle of 0°. However, at a solar zenith angle of 60°, the porous cloud had a value of cloud absorptance greater than its homogeneous counterpart. This behavior is attributed to the ability of incident energy to penetrate deeper into the cloud via the columns before scattering; consequently the energy has a smaller chance of scattering back through the cloud top. When pockets of clear air were introduced into an infinite cloud layer almost no change was observed in the bulk radiative properties.

Clouds in which microstructure has a high degree of vertical and horizontal inhomogeneity were also studied. This was achieved by randomly assigning values of droplet concentration (between 0 and 100 cm^{-3}) in 243 equal volume boxes of finite clouds and repeated arrays making up infinite clouds. It was found that bulk radiative parameters were relatively insensitive to different randomized distributions. Additionally, when compared to homogeneous clouds with equal mean droplet concentrations, randomized clouds showed similar values of the bulk radiative parameters. Small differences in reflectance and transmittance were noted at a 0° zenith angle due to a channeling effect which occurs in the randomized models. This random alignment of smaller density elements results in slightly greater transmittance and smaller reflectance in the random model relative to the mean homogeneous model.

As would be expected, the introduction of columns of large (rain) droplets into finite and infinite clouds resulted in bulk radiative parameters whose values fall between those corresponding to homogeneous clouds and clouds with columns of clear air.

Columns of bimodally distributed droplets were introduced into plane-parallel clouds. As the small-droplet concentration was varied according to the small-droplet plus large-droplet progression C.5 + Rain 50, C.5/10 + Rain 50 and C.5/100 + Rain 50, a transition was observed in the bulk radiative properties from those more characteristic of a homogeneous C.5 cloud (approximated by the C.5 + Rain 50 cores) to those of a cloud with a monomodal Rain 50 core (approximated by the C.5/100 + Rain 50 cores).

REFERENCES

Aida, M., 1976: Transfer of solar radiation in an array of cumuli. Paper delivered at Symposium on Radiation in the Atmosphere, Garmisch-Partenkirchen, Germany, 19–28 August.

——, 1977: Scattering of solar radiation as a function of cloud dimensions and orientation. *J. Quant. Spectrosc. Radiat. Transfer,* **17,** 303–310.

Albrecht, B., 1979: A model of the thermodynamic structure of the trade-wind boundary layer: Part II. Applications. *J. Atmos. Sci.,* **36,** 90–98.

——, and S. K. Cox, 1975: The large-scale response of the tropical atmosphere to cloud-modulated infrared heating. *J. Atmos. Sci.,* **32,** 16–24.

Appleman, H. S., 1961: Occurrence and forecasting of cirrostratus clouds. World Meteorological Organization No. 109–47, Tech. Note No. 40.

Barkstrom, B. R., and R. F. Arduini, 1976: The effect of finite horizontal size of clouds upon the visual albedo of the earth. Paper delivered at Symposium on Radiation in the Atmosphere, Garmisch-Partenkirchen, Germany, 19–28 August.

Bertie, J. E., H. J. Labbe and E. Whalley, 1969: Absorptivity of ice in the range 400–30 cm^{-1}. *J. Chem. Phys.,* **50,** 4501–4520.

Best, A. C., 1951: Drop-size distribution in clouds and fog. *Quart. J. Roy. Meteor. Soc.,* **77,** 418–426.

Braham, R. R., Jr., and P. Spyers-Duran, 1967: Survival of cirrus crystals in clean air. *J. Appl. Meteor.,* **6,** 1053–1061.

Busygin, V. P., N. A. Yevstratov and E. M. Feigelson, 1973: Optical properties of cumulus clouds and radiant fluxes for cumulus cloud cover. *Izv. Atmos. Ocean. Phys.,* **9,** 1142–1151.

Chýlek, P., 1975: Asymptotic limits of the Mie-scattering characteristics. *J. Opt. Soc. Amer.,* **65,** 1316–1317.

——, 1976: Partial-wave resonances and the ripple structure in the Mie normalized extinction cross-section. *J. Opt. Soc. Amer.,* **66,** 285–287.

——, 1977: A note on extinction and scattering efficiencies. *J. Appl. Meteor.,* **16,** 321–322.

——, G. W. Grams and R. G. Pinnick, 1976: Light scattering by irregular randomly oriented particles. *Science,* **193,** 480–482.

Cox, S. K., 1969: Radiation models of midlatitude synoptic features. *Mon. Wea. Rev.,* **97,** 637–651.

——, T. H. Vonder Haar and V. Suomi, 1973: Measurements of absorbed shortwave energy in a tropical atmosphere. *Solar Energy,* **14,** 169–173.

Crane, R. K., 1977: Predictions of the effects of rain on satellite communication systems. *Proc. IEEE,* **65,** 456–474.

Davies, R., 1976: The three-dimensional transfer of solar radiation in clouds. Ph.D. thesis. University of Wisconsin, Madison, 220 pp.

——, 1978: The effect of finite geometry on the three-dimensional transfer of solar irradiance in clouds. *J. Atmos. Sci.,* **35,** 1712–1725.

Davis, J. M., S. K. Cox and T. B. McKee, 1979a: Total shortwave radiative characteristics of absorbing finite clouds. *J. Atmos. Sci.,* **36,** 508–518.

——, —— and ——, 1979b: Vertical and horizontal distributions of solar absorption in finite clouds. *J. Atmos. Sci.,* **36,** 1976–1984.

de Almeida, F. C., 1977: Collision efficiency, collision angle and impact velocity of hydrodynamically interacting cloud drops: A numerical study. *J. Atmos. Sci.,* **34,** 1286–1292.

Deirmendjian, D., 1969: *Electromagnetic Scattering on Spherical Polydispersions.* Elsevier, 290 pp.

——, 1975: Far-infrared and sub-millimeter wave attenuation by clouds and rain. *J. Appl. Meteor.,* **14,** 1584–1593.

Donn, B., and R. S. Powell, 1963: *Electromagnetic Scattering,* M. Kerker, Ed. Pergamon Press, 287 pp.

Feigelson, E. M., and L. D. Krasnokutskaya, 1978: *Fluxes of Solar Radiation in Clouds.* Hydrometeorological Publishing House, Leningrad.

Fingerhut, W. A., 1977: A numerical model of a diurnally varying tropical cloud cluster disturbance. Paper prepared for U.S. Workshop on the GATE Central Program, NCAR, Boulder, 25 July–12 August, 39 pp.

Fleming, J. R., and S. K. Cox, 1974: Radiative effects of cirrus clouds. *J. Atmos. Sci.,* **31,** 2182–2188.

Foltz, G. S. and W. M. Gray, 1979: Diurnal variation in the troposphere's energy balance. *J. Atmos. Sci.,* **36,** 1450–1466.

Frank, W. M., 1977: The life-cycle of GATE convective systems. Paper prepared for U.S. Workshop on GATE Central Program, NCAR, Boulder, 25 July–12 August, 37 pp.

Gifford, M. D., and T. B. McKee, 1977: Characteristic size spectra of cumulus fields observed from satellites. Atmos. Sci. Pap. 280, Colorado State University, 90 pp. [NTIS PB292997/AS].

Gille, J. C., and T. N. Krishnamurti, 1972: On radiative interactions in a tropical disturbance. *Preprints Conf. Atmospheric Radiation,* Ft. Collins, Amer. Meteor. Soc., 266–268.

Gray, W. M., and R. Jacobson, 1977: Diurnal variation of deep cumulus convection. *Mon. Wea. Rev.,* **105,** 1171–1188.

Griffith, K. T., S. K. Cox and R. G. Knollenberg, 1980: Infrared radiative properties of tropical cirrus clouds inferred from aircraft measurements. *J. Atmos. Sci.,* **37** (in press).

Grube, P. B., 1977: Influence of deep cumulus convection on upper tropospheric temperature changes in GATE. Paper prepared for U.S. Workshop on the GATE Central Program, NCAR, Boulder, 25 July–12 August, 18 pp.

Hansen, J. E., 1971: Multiple scattering of polarized light in planetary atmospheres. Part II. Sunlight reflected by terrestrial water clouds. *J. Atmos. Sci.,* **28,** 1400–1426.

Heymsfield, A., 1972: Ice crystal terminal velocities. *J. Atmos. Sci.,* **29,** 1348–1357.

——, 1975: Cirrus uncinus generating cells and the evolution of cirriform clouds. Part I: Aircraft observations of the growth of the ice phase. *J. Atmos. Sci.,* **32,** 799–808.

——, 1977: Precipitation development in stratiform ice clouds: A microphysical and dynamical study. *J. Atmos. Sci.,* **34,** 367–381.

——, and R. G. Knollenberg, 1972: Properties of cirrus generating cells. *J. Atmos. Sci.,* **29,** 1350–1366.

Hobbs, P. V., 1974: High concentrations of ice particles in a larger cloud. *Nature,* **251,** 694–696.

Hodkinson, J. R., 1963: *Electromagnetic Scattering,* M. Kerker, Ed. Pergamon Press, 287 pp.

——, 1966: *Aerosol Science,* C. N. Davies, Ed. Academic Press, 287 pp.

Hogg, D. C., and T. S. Chu, 1975: The role of rain in satellite communications. *Proc. IEEE,* **63,** 1308–1329.

Holland, A. C., and J. S. Draper, 1967: Analytical and experimental investigation of light scattering from polydispersions of Mie particles. *Appl. Opt.,* **6,** 511–518.

——, and G. Gagne, 1970: The scattering of polarized light by polydisperse systems of irregular particles. *Appl. Opt.,* **9,** 1113–1121.

Holton, J., 1971: A diagnostic model for equatorial disturbances: The role of vertical shear from the mean zonal wind. *J. Atmos. Sci.,* **28,** 55–64.

Howard, J. N., D. L. Burch and D. Williams, 1955: Near-infrared transmission through synthetic atmospheres. *J. Opt. Soc. Amer.*, **46,** 186–190.

Hunt, G. E., 1973: Radiative properties of terrestrial clouds at visible and infrared thermal window wavelengths. *Quart. J. Roy. Meteor. Soc.*, **99,** 346–369.

Inada, H., 1974: Backscattered short pulse response of surface waves from dielectric spheres. *Appl. Opt.*, **13,** 1928–1933.

Irvine, W. M., and J. B. Pollack, 1968: Infrared optical properties of water and ice spheres. *Icarus*, **8,** 324–366.

Jacobowitz, H., 1970: Emission, scattering and absorption of radiation in cirrus cloud layers. Ph.D. dissertation, M.I.T., 181 pp.

——, 1971: A method for computing the transfer of solar radiation through clouds of hexagonal ice crystals. *J. Quant. Spectros. Radiat. Transfer*, **11,** 691–695.

Jonas, P. R., and P. Goldsmith, 1972: The collection efficiencies of small droplets falling through a sheared flow. *J. Fluid Mech.*, **52,** 593–608.

——, and B. J. Mason, 1974: The evolution of droplet spectra by condensation and coalescence in cumulus clouds. *Quart. J. Roy. Meteor. Soc.*, **100,** 286–295.

Jones, Douglas M. A., 1959: The shape of raindrops. *J. Meteor.*, **16,** 504–510.

Joseph, J., W. J. Wiscombe and J. A. Weinman, 1976: The delta-Eddington approximation for radiative flux transfer. *J. Atmos. Sci.*, **33,** 2452–2459.

Knollenberg, R. G., 1975: The response of optical array spectrometers to ice and snow: A study of probe size to crystal mass relationships. Air Force Cambridge Res. Labs., AFCRL-TR-75-0494.

Kuenning, J. A., T. B. McKee and S. K. Cox, 1978: A laboratory investigation of radiative transfer in cloud fields. Atmos. Sci. Pap. 286, Colorado State University, Fort Collins, 66 pp. [NTIS PB293018/AS].

Lilly, D. K., 1968: Models of cloud-topped mixed layers under a strong inversion. *Quart. J. Roy. Meteor. Soc.*, **94,** 292–309.

Linkes Meteorologisches Taschenbuch II, 1953: Franz Baur, Leipzig Germany Akademische Verlagsgesellschaft Geest and Portig K-G.

Liou, K. N., 1972a: Light scattering by ice clouds in the visible and infrared: A theoretical study. *J. Atmos. Sci.*, **29,** 524–536.

——, 1972b: Electromagnetic scattering by arbitrarily oriented ice cylinders. *Appl. Opt.*, **11,** 667–674.

——, 1974: On the radiative properties of cirrus in the window region and their influence on remote sensing of the atmosphere. *J. Atmos. Sci.*, **31,** 522–532.

——, and T. Sasamori, 1975: On the transfer of solar radiation in aerosol atmospheres. *J. Atmos. Sci.*, **32,** 2166–2177.

Lopez, R. E., 1976: Radar characteristics of the cloud populations of tropical disturbances in the northwest Atlantic. *Mon. Wea. Rev.*, **104,** 268–283.

Manabe, S., and R. F. Strickler, 1964: Thermal equilibrium of the atmosphere with convective adjustment. *J. Atmos. Sci.*, **21,** 261–385.

Marshall, J. S., and W. M. Palmer, 1947: The distribution of raindrops with size. *J. Meteor.*, **5,** 165–166.

Mason, B. J., 1971: *The Physics of Clouds*. Clarendon Press, 671 pp.

——, and P. R. Jonas, 1974: The evolution of droplet spectra and large droplets by condensation in cumulus clouds. *Quart. J. Roy. Meteor. Soc.*, **100,** 23–38.

McClatchey, R. A., R. W. Fenn, J. E. A. Selby, E. E. Volz and J. S. Garing, 1971: Optical properties of the atmosphere (revised). AFCRL *Environ. Res. Pap.*, No. 354, Bedford, MA.

McKee, T. B., and S. K. Cox, 1974: Scattering of visible radiation by finite clouds. *J. Atmos. Sci.*, **31,** 1885–1892.

——, and ——, 1976: Simulated radiance patterns for finite cubic clouds. *J. Atmos. Sci.*, **33,** 2014–2020.

——, and J. T. Klehr, 1976: Radiative effects of cloud geometry. *Proceedings of Symposium on Radiation in the Atmosphere*, Garmisch-Partenkirchen, 217–219.

McTaggart-Cowan, J. D., G. Lala and B. Vonnegut, 1970: The design, construction and use of an ice crystal counter for ice crystal studies by aircraft. *J. Appl. Meteor.*, **9,** 294–299.

Medhurst, R. G., 1965: Rainfall attenuation of centimeter waves: comparison of theory and measurement. *IEEE Trans. Antennas Propag.*, **AP-13,** 550–563.

Möller, F., 1943: Labilisierung von schichtwolken durch strahlung. *Meteor. Z.*, **60,** 212–213.

Napper, D. H., and R. H. Ottewill, 1963: In *Electromagnetic Scattering*. M. Kerker. Ed. Pergamon Press, 377 pp.

Oguchi, T., 1975: Rain depolarization studies at centimeter and millimeter wavelengths: theory and measurement. *J. Radio Res. Lab. Japan*, **22,** 165–211.

Okita, J., 1961: Size distribution of large droplets in precipitating clouds. *Tellus*, **13,** 509–521.

Oliver, D. A., W. S. Lewellen and G. G. Williamson, 1978: The interaction between turbulent and radiative transport in the development of fog and low-level stratus. *J. Atmos. Sci.*, **35,** 301–316.

Ono, A., 1969: The shape and riming properties of ice crystals in natural clouds. *J. Atmos. Sci.*, **26,** 138–147.

Paltridge, G. W., 1974a: Atmospheric radiation and the gross character of stratiform clouds. *J. Atmos. Sci.*, **31,** 244–250.

——, 1974b: Global cloud cover and earth surface temperature. *J. Atmos. Sci.*, **31,** 1571–1576.

Pinnick, R. G., D. E. Carrol and D. J. Hofman, 1976: Polarized light scattered from monodisperse randomly oriented nonspherical aerosol particles: measurements. *Appl. Opt.*, **15,** 384–393.

Plank, V. G., 1969: The size distribution of cumulus clouds in representative Florida populations. *J. Appl. Meteor.*, **8,** 46–67.

Platt, C. M. R., 1976: Infrared absorption and liquid water content in stratocumulus clouds. *Quart. J. Roy. Meteor. Soc.*, **102,** 515–522.

——, 1977: Lidar observations of a mixed-phase altostratus cloud. *J. Appl. Meteor.*, **16,** 339–345.

——, 1978: Lidar backscatter from horizontal ice crystal plates. *J. Appl. Meteor.*, **17,** 482–488.

Poellot, M. R., and S. K. Cox, 1977: Computer simulation of irradiance measurements from aircraft. *J. Appl. Meteor.*, **16,** 167–171.

Pruppacher, H. R., and R. L. Pitter, 1971: A semi-empirical determination of the shape of cloud and rain drops. *J. Atmos. Sci.*, **28,** 86–94.

——, and J. D. Klett, 1978: *Microphysics of Atmospheric Clouds and Precipitation*. D. Reidel Publishing Co., Dordrecht, Holland.

Ray, P. S., and J. J. Stephens, 1974: Far-field transient backscattering by ice spheres. *Radio Sci.*, **9,** 43–55.

Reynolds, D. W., and T. H. Vonder Haar, 1973: A comparison of radar-determined cloud height and reflected solar radiance measured from the geosynchronous satellite ATS-3. *J. Appl. Meteor.*, **12,** 1082–1084.

——, —— and S. K. Cox, 1975: The effect of solar radiation in the tropical troposphere. *J. Appl. Meteor.*, **14,** 433–444.

——, T. B. McKee and K. S. Danielson, 1978: Effects of cloud size and cloud particles on satellite-observed reflected brightness. *J. Atmos. Sci.*, **35,** 160–164.

Roach, W. T., 1961: Some aircraft observations of fluxes of solar radiation in the atmosphere. *Quart. J. Roy. Meteor. Soc.*, **87,** 346–363.

Rogers, R. R., 1976: Statistical rainstorm models: their theoretical and physical foundations. *IEEE Trans. Antennas Propag.*, **AP-24,** 547–566.

Rosinski, J., C. T. Nagamoto, G. Langer and F. Parungo, 1970: Cirrus clouds as collectors of aerosol particles. *J. Geophys. Res.*, **75,** 2961–2973.

Ryan, R. T., H. Blau Jr., P. C. Von Thüna, M. L. Cohen and C. D. Roberts, 1972: Cloud microstructure as determined by an optical cloud particle spectrometer. *J. Appl. Meteor.*, **11,** 149–156.

Sartor, J. D., and T. W. Cannon, 1977: Collating airborne and surface observations of the microstructure of precipitating continental convective clouds. *J. Appl. Meteor.*, **16,** 697–707.

Schaaf, J. W., and D. Williams, 1973: Optical constants of ice in the infrared. *J. Opt. Soc. Amer.*, **63,** 726–732.

Schubert, W. H., 1976: Experiments with Lilly's cloud-topped mixed layer model. *J. Atmos. Sci.*, **33,** 436–446.

——, J. S. Wakefield, E. J. Steiner and S. K. Cox, 1977: Marine stratocumulus convection. Atmos. Sci. Pap. 273, Colorado State University, 140 pp. [NTIS PB 272955/AS].

Semplak, R. A., 1970: The influence of heavy rainfall on attenuation at 18.5 and 30.9 GHz. *IEEE Trans. Antennas Propag.*, **AP-18,** 507–511.

Smithsonian Meteorological Tables, 1966: Smithsonian Institution, 527 pp.

Squires, P., 1958: Penetrative downdrafts in cumuli. *Tellus*, **10,** 381–385.

Stephens, G. L., 1976: The transfer of radiation through vertically non-uniform stratocumulus water clouds. *Beitr. Phys. Atmos.*, **49,** 237–253.

——, 1978: Radiation profiles in extended water clouds. I: Theory. *J. Atmos. Sci.*, **35,** 2111–2122.

Stephens, J. J., P. S. Ray and R. J. Kurzeja, 1971: Far-field transient backscattering by water drops. *J. Atmos. Sci.*, **28,** 785–793.

Strand, K. A., 1963: *Basic Astronomical Data*. The University of Chicago Press, 495 pp.

Tampieri, F., and C. Tomasi, 1976: Size distribution models of fog and cloud droplets in terms of the modified gamma function. *Tellus*, **28,** 333–347.

Tennekes, H., and J. D. Woods, 1973: Coalescence in a weakly turbulent cloud. *Quart. J. Roy. Meteor. Soc.*, **99,** 758–763.

Tverskoi, P. N., 1965: *Physics of the Atmosphere*. Israel Program for Scientific Translations, Jerusalem, 314–343.

Twomey, S., 1972: The effect of cloud scattering on the absorption of solar radiation by atmospheric dust. *J. Atmos. Sci.*, **29,** 1156–1159.

——, 1976: Computations of the absorption of solar radiation by clouds. *J. Atmos. Sci.*, **33,** 1087–1091.

Van Blerkom, D. J., 1971: Diffuse reflection from clouds with horizontal inhomogeneities. *Astrophys. J.*, **166,** 235–242.

Van de Hulst, H. C., 1957: *Light Scattering by Small Particles*. Wiley, 470 pp.

Varley, D. J., 1978a: Cirrus particle distribution study. Part 1. AFGL-TR-78-0192, 71 pp.

——, 1978b: Cirrus particle distribution study, Part 3. AFGL-TR-78-0305, 67 pp.

——, and D. M. Brooks, 1978: Cirrus particle distribution study, Part 2. AFGL-TR-78-0248, 108 pp.

——, and A. Barnes, 1979: Cirrus particle distribution study, Part 4. AFGL-TR-79-0134, 91 pp.

Waldvogel, A., 1974: The N_0 jump of raindrop spectra. *J. Atmos. Sci.*, **31,** 1067–1075.

Warner, J., 1973: The microstructure of cumulus clouds. Part V. Changes in droplet size distribution with cloud age. *J. Atmos. Sci.*, **30,** 1724–1726.

——, 1977: Time variation of updraft and water content of small cumulus clouds. *J. Atmos. Sci.*, **34,** 1306–1312.

Weickmann, H. K., 1947: Die Eisphase in der Atmosphere. Library Trans. 273, Royal Aircraft Establishment, Farnborough, 96 pp.

Welch, R. M., J. F. Geleyn, W. G. Zdunkowski and G. Korb, 1976: Radiation transfer of solar radiation in model clouds. *Beitr. Phys. Atmos.*, **49,** 128–146.

——, and S. K. Cox, 1978: Nonspherical extinction and absorption efficiencies. *Appl. Opt.*, **17,** 3159–3168.

Wendling, P., 1977: Albedo and reflected radiance of horizontally inhomogeneous clouds. *J. Atmos. Sci.*, **34,** 642–650.

Wiscombe, W. J., 1976: On initialization, error and flux conservation in the doubling method. *J. Quant. Spectros. Radiat. Transfer*, **16,** 637–658.

——, 1977: The delta-M method: rapid yet accurate radiative flux calculations for strongly asymmetric phase functions. *J. Atmos. Sci.*, **34,** 1408–1422.

——, and J. W. Evans, 1977: Exponential-sum fitting of radiative transmission functions. *J. Comput. Phys.*, **24,** 416–444.

Yamamoto, G., 1962: Direct absorption of solar radiation by atmospheric water vapor, carbon dioxide and molecular oxygen. *J. Atmos. Sci.*, **19,** 182–188.

Yanai, M., S. Esbensen and J. Chu, 1973: Determination of the bulk properties of tropical cloud clusters from large-scale heat and moisture budgets. *J. Atmos. Sci.*, **30,** 611–627.

Zdunkowski, W. G., B. C. Nielsen, and G. Korb, 1967: Prediction and maintenance of radiation fog. Tech. Rep. ECOM-0049-F, USAEC, Fort Monmouth, N.J.

——, and C. T. Davis Jr., 1974: Radiative transfer in vertically inhomogeneous layer clouds. *Beitr. Phys. Atmos.*, **47,** 187–212.

——, and G. Korb, 1974: An approximative method for the determination of short-wave radiative fluxes in scattering and absorbing media. *Beitr. Phys. Atmos.*, **47,** 129–144.

——, and J. D. Pryce, 1974: The approximate distribution of scattered solar radiative intensities from optically thin cirrus. *Pure Appl. Geophys.*, **112,** 739–752.

Zerull, R. H., 1976: Scattering measurements of dielectric and absorbing non-spherical particles. *Beitr. Phys. Atmos.*, **49,** 168–188.

SUBJECT INDEX

When t follows the page number it signifies a table.
When f follows the page number it signifies a figure.
When d follows the page number it signifies a definition.